Lecture Note
Mathematics

A collection of informal reports and seminars
Edited by A. Dold, Heidelberg and B. Eckmann, Zürich

Series: Tata Institute of Fundamental Research, Bombay
Adviser: M. S. Narasimhan

286

Shingo Murakami

Osaka University, Toyonaka, Osaka/Japan

On Automorphisms of Siegel Domains

Springer-Verlag
Berlin · Heidelberg · New York 1972

AMS Subject Classifications (1970): 32 A 07, 32 M xx

ISBN 3-540-05985-7 Springer-Verlag Berlin · Heidelberg · New York
ISBN 0-387-05985-7 Springer-Verlag New York · Heidelberg · Berlin

Offsetdruck: Julius Beltz, Hemsbach/Bergstr.

PREFACE

These are notes of the lectures which I gave at the Tata Institute of Fundamental Research in the Winter 1971.

In these lectures I intended to introduce recent results on the automorphisms of Siegel Domains in a unified form. Materials of these notes are the papers listed at the end of Introduction. Other necessary references are given on occasion in the notes.

My thanks are due to Mr. and Mrs. Dani for their preparation of these notes, and to M.S. Narasimhan and M.S. Raghunathan for their helpful suggestions.

S. MURAKAMI

C O N T E N T S

Introduction

We denote by $\mathbb{C}^N$ the N dimensional complex vector space along with its usual analytic structure. An open connected subset of $\mathbb{C}^N$ is called a domain. For a domain D, $G(D)$ denotes the group of all holomorphic automorphisms of D.

We note the following facts.

I If D is a bounded domain, then D has a volume element (called the Bergman volume) v, which is invariant under the action of $G(D)$. Let

$$v = i^{N^2} K dz^1 \wedge \dots \wedge dz^N \wedge d\bar{z}^1 \wedge \dots \wedge d\bar{z}^N$$

where $z^1, z^2, \dots z^N$ are complex coordinates in $\mathbb{C}^N$ and K is a positive function. In this case

$$ds^2 = \sum_{h,j} \frac{\partial^2 \log K}{\partial z^h \partial \bar{z}^j} dz^h \wedge d\bar{z}^j ;$$

is a Kaehler metric invariant under the action of $G(D)$.

II The group $G(D)$ has a structure of real Lie group and with this structure $G(D)$ acts on D differentiably.

Definition. A domain D is said to be homogeneous if the group $G(D)$ acts transitively on D.

In this case the connected component of the identity of $G(D)$ acts transitively on D.

Examples

1) Let D be the unit disc $\{z \in \mathbb{C} \mid |z| < 1\}$. Then

$$G(D) = \{kg_\alpha \mid \alpha \in D,\ k \in \mathbb{C} \text{ and } |k| = 1\},$$

where $g_\alpha : D \longrightarrow D$ is defined by

$$g_\alpha(z) = \frac{z - \alpha}{\bar{\alpha} z - 1}$$

Clearly $G(D)$ acts transitively on D and therefore D is homogeneous.

2) Let $N = 2$. Consider the following domains

$$D_1 = \left\{(z^1, z^2) \in \mathbb{C}^2 \,\middle|\, |z^1|^2 + |z^2|^2 < 1\right\},$$

$$D_2 = \left\{(z^1, z^2) \in \mathbb{C}^2 \,\middle|\, \operatorname{Sup}\{|z^1|, |z^2|\} < 1\right\}.$$

It is checked that D_1 and D_2 are homogeneous domains. It is known that they are non-isomorphic and that any homogeneous bounded domain in $\mathbb{C}^2$ is holomorphically isomorphic to D_1 or D_2.

Definition. A domain D is said to be symmetric if for any $z \in D$ there exists $\sigma_z \in G(D)$ such that σ_z^2 = Identity and that σ_z has z as an isolated fixed point.

It is easily proved that a symmetric bounded domain is homogeneous. For a homogeneous domain to be symmetric it is sufficient that the above condition be satisfied at one point.

Examples. D (Unit disc in the plane), D_1, D_2 are all symmetric domains. In fact $\sigma_0(z) = -z$ is the symmetry at the origin ($z \in D$, D_1 or D_2).

E. Cartan gave a complete classification of symmetric bounded domains in $\mathbb{C}^N$, and he asked whether any bounded homogeneous domain is symmetric or not (1935). In this connection following things are now known.

i) This is true if the dimension N of the domain ≤ 3. (Cartan 1935).

ii) This is true if there exists a semisimple Lie group acting transitively on D. (Koszul, Borel 1954).

iii) This is true if there exists a unimodular Lie group acting transitively on D. (Hano 1957).

iv) However in 1957, Pjateckii-Shapiro found a counter example to the Cartan's problem.

We also note here the following

Theorem. (Vinberg, Gindinkin and Pjateckii-Shapiro): If D is a bounded homogeneous domain in $\mathbb{C}^N$, then there exists an (unbounded) Siegel domain D' in $\mathbb{C}^N$ which is holomorphically isomorphic to D; moreover a group of affine automorphisms of $\mathbb{C}^N$ acts transitively on D.

The meaning of Siegel domain in the above theorem will be explained later.

Example. The isomorphism of the unit disc D and the upper half plane H (which is a Siegel domain) is an example of the above theorem. The isomorphism $D \longrightarrow H$ is given by $z \longrightarrow i\,\frac{1+z}{1-z}$ and the group of affine transformations $z \longrightarrow az + b$ $(a > 0,\ b \in \mathbb{R})$ acts transitively on the upper half-plane.

Now the main subjects treated in these lectures are connected with the structure of the group G(D) for a Siegel domain D. As applications of the results, we will get certain sufficient conditions for D to be symmetric or to be homogeneous.

Materials of these lectures are found in the papers:

[1] W. Kanp, Y. Matsushima, T. Ochiai: On automorphisms and equivalences of generalised Siegel domains. American J. Math. 92 (1970) 475-497.

[2] N. Tanaka: On the automorphisms of Siegel domains, J. Math. Soc. Japan 22 (1970) 180-217.

[3] J. Vey: Sur la division des domains de Siegel, Ann. Ecole Normale Superieure 4 ieme Serie 3 (1971) 479-506.

[4] S.G. Gindinkin, I.I. Pjatecckii-Šapiro, E.B. Vinberg: Homogeneous Kǎhler manifolds, "Geometry of Homogeneous Bounded Domains" C.I.M.E. 1967, 1-88.

For the definition and elementary properties on Siegel domains, we follow Pjatecckii-Šapiro, but apart from these things we shall not be concerned with the theory of Kaehler-algebra as developed by Gindinkin,Pjatecckii-Sapiro and Vinberg. (e.g. [4]). While, main results of [2] will be derived in these lectures directly from those of [1] in a rather simple way.

§ 1. Siegel Domains

Definition. A subset Ω of $\mathbb{R}^n$ is said to be a cone if for all $x \in \Omega$ and $\lambda > 0$, λx is also in Ω. A cone Ω is said to be convex if for any $x, y \in \Omega$ all points of the form $\lambda x + \mu y$ belong to Ω where $\lambda \geqslant 0,\ \mu \geqslant 0,\ \lambda + \mu = 1$.

In the following discussion let Ω denote an open convex cone in $\mathbb{R}^n$ not containing any straight line.

<u>Definition</u>. The set $D(\Omega) = \mathbb{R}^n + i\Omega$

$$= \{x + iy \mid x \in \mathbb{R}^n,\ y \in \Omega\}$$

in $\mathbb{C}^n$ is called a <u>Siegel domain of the first kind</u> associated to Ω , or a <u>tube domain over</u> Ω .

<u>Example</u>. Let $n = 1$ and $\Omega = \mathbb{R}^+ = \{x \in \mathbb{R} \mid x > 0\}$. Then $D(\mathbb{R}^+)$ is the upper half plane in $\mathbb{C}^1$.

<u>Proposition 1.1</u>. Let $D(\Omega) \subset \mathbb{C}^n$ be a Siegel domain of the first kind. Then $D(\Omega)$ is isomorphic to a bounded domain in $\mathbb{C}^n$.

<u>Proof</u>. It is obvious that $D(\Omega)$ is a domain in $\mathbb{C}^n$. Let $\Omega' = \{(x^1,\ldots,x^n) \in \mathbb{R}^n \mid x^i > 0 \text{ for all } i\}$ be the positive cone in $\mathbb{R}^n$. Since Ω does not contain any straight line, up to a linear transformation we may assume that $\Omega \subset \Omega'$. Thus $D(\Omega)$ is contained in $D(\Omega')$ as an open subset; and hence it is sufficient to prove that $D(\Omega')$ is isomorphic to a bounded domain. Now

$$\begin{aligned} D(\Omega') &= \mathbb{R}^n + i\Omega' \\ &= \{(z^1,\ldots,z^n) \in \mathbb{C}^n \mid \operatorname{Im} z^i > 0 \text{ for all } i\}. \\ &= D^1 \times D^2 \times \ldots \times D^n \end{aligned}$$

where $D^i = \{z^i \in \mathbb{C} \mid \operatorname{Im} z^i > 0\}$ $i = 1,2,\ldots n$ are copies of upper half plane. But D^i is isomorphic to the unit disc B. Therefore $D(\Omega')$ is isomorphic to the product $B \times B \times \ldots \times B$ of n copies of B. Thus $D(\Omega')$ is isomorphic to a bounded domain.

(Q.E.D.)

<u>Example</u>. Let $H(n,\mathbb{R})$ denote the set of real symmetric matrices of degree n. Then $H(n,\mathbb{R})$ is a real vector space of dimension $\frac{1}{2}n(n+1)$. Let $H^+(n,\mathbb{R})$ be the set of all positive definite real symmetric matrices of degree n. It is

known that a real symmetric matrix is positive definite if and only if all its eigenvalues are positive. From this it can be proved that $H^+(n,\mathbb{R})$ is an open subset of $H(n,\mathbb{R})$. It is easily seen that $H^+(n,\mathbb{R})$ is a convex cone not containing any straight line in $H(n,\mathbb{R})$. Take $H^+(n,\mathbb{R})$ for Ω. Then

$$D(\Omega) = H(n,\mathbb{R}) + i\, H^+(n,\mathbb{R})$$
$$= \left\{ z \in H(n,\mathbb{C}) \;\middle|\; \operatorname{Im} z > 0 \right\}$$

canonically, where $H(n,\mathbb{C})$ is the set of all complex symmetric matrices of degree n and > 0 denotes positive-definiteness. This Siegel domain is called <u>Siegel's upper half space</u>. We prove that $D(\Omega)$ is isomorphic to the domain $D = \left\{ z \in H(n,\mathbb{C}) \;\middle|\; 1_n - {}^t z\, \bar{z} > 0 \right\}$ bounded in $H(n,\mathbb{C})$, where 1_n is the unit matrix of degree n.

<u>Proof</u>. Define $\rho : D(\Omega) \longrightarrow D$ by $\rho(z) = (z - i1_n)(z + i1_n)^{-1}$.

If $z \in D(\Omega)$, we have

$$(z + i1_n)\,{}^t\overline{(z + i1_n)} = ({z}\,{}^t\bar{z} + 1_n) + 2 \operatorname{Im} z > 0.$$

So $(z + i1_n)$ is invertible, since a positive definite matrix is invertible. Further it is easy to show that $w = (z - i1_n)(z + i1_n)^{-1}$ is in D for $z \in D(\Omega)$.

Similarly we can define $\sigma : D \longrightarrow D(\Omega)$ by $\sigma(w) = i(w + 1_n)(1_n - w)^{-1}$. It is seen that both these maps are holomorphic and are inverses of each other.

We now come to another kind of domains called Siegel domains of the second kind. For motivation consider the following example. Put

$$B^N = \left\{ (z^1,\ldots,z^N) \in \mathbb{C}^N \;\middle|\; \sum_1^N |z^k|^2 < 1 \right\},$$

$$\mathcal{E}^N = \left\{ (z, w^1, w^2, \ldots w^{N-1}) \in \mathbb{C}^1 \times \mathbb{C}^{N-1} \;\middle|\; \operatorname{Im} z - \sum_1^{N-1} |w^k|^2 > 0 \right\}.$$

We prove that B^N is holomorphically isomorphic to $\mathcal{E}^N$. Define a map ρ of $\mathbb{C}^N$ onto $\mathbb{C}^N$ by $(z, w^1, w^2 \ldots w^{N-1}) \longrightarrow (z^1, z^2 \ldots z^N)$ where

$$z^1 = \frac{z-i}{z+i}, \quad z^2 = \frac{2w^1}{z+i}, \ldots, \quad z^N = \frac{2w^{N-1}}{z+i},$$

namely

$$z = i\,\frac{1+z^1}{1-z^1}\ ,\ w^1 = i\,\frac{z^2}{1-z^1}\ ,\dots\ w^{N-1} = i\,\frac{z^N}{1-z^1}$$

Then we have

$$1 - \sum_{k=1}^{N} |z^k|^2 = \frac{4}{|z+i|^2}\left\{\operatorname{Im} z - \sum_{k=1}^{N-1} |w^k|^2\right\},$$

and it follows that the above map gives rise to an isomorphism of $\mathcal{E}^N$ onto B^N. We shall call $\mathcal{E}^N$ the <u>elementary Siegel domain</u> of dimension N, and we make the convention that $\mathcal{E}^1$ is the upper half plane in $\mathbb{C}^1$.

<u>Definition</u>. Let $\Omega \subset \mathbb{R}^n$ be an open convex cone not containing any straight line. A map $F : \mathbb{C}^m \times \mathbb{C}^m \longrightarrow \mathbb{C}^n$ is said to be an Ω -<u>hermitian form</u> if it satisfies the following conditions.

i) For each $w' \in \mathbb{C}^m$ the map $F_{w'} : \mathbb{C}^m \longrightarrow \mathbb{C}^n$ defined by $F_{w'}(w) = F(w, w')$ is complex linear.

ii) $F(w',w) = \overline{F(w,w')}$, where $\bar{z}$ denotes the conjugate of $z \in \mathbb{C}^n$ with respect to $\mathbb{R}^n$.

iii) $F(w,w)$ belongs to the closure $\bar{\Omega}$ of Ω for any $w \in \mathbb{C}^m$.

iv) $F(w,w) = 0$ if and only if $w = 0$.

If Ω is the set $\mathbb{R}^+$ of positive real numbers, an Ω -hermitian form reduces to a positive definite hermitian form.

<u>Definition</u>. Let $\Omega \subset \mathbb{R}^n$ be an open convex cone not containing any straight line and $F : \mathbb{C}^m \times \mathbb{C}^m \longrightarrow \mathbb{C}^n$ be an Ω -hermitian form. Then the set

$$D(\Omega,F) = \left\{(z,w) \mid z \in \mathbb{C}^n,\ w \in \mathbb{C}^m,\ \operatorname{Im} z - F(w,w) \in \Omega\right\}$$

is called <u>a Siegel domain of the second kind associated to</u> (Ω,F).

An elementary Siegel domain $\mathcal{E}^N$ is a Siegel domain of the second kind associated to (Ω,F) where $\Omega = \mathbb{R}^+$ and $F(w,w') = \sum_{k=1}^{N-1} w^k \bar{w}'^k$ for $w,w' \in \mathbb{C}^{N-1}$.

Note that $D(\Omega,F)$ is a domain in $\mathbb{C}^{n+m}$. In fact let $\alpha : \mathbb{C}^n \times \mathbb{C}^m \longrightarrow \mathbb{R}^n \times \mathbb{R}^n \times \mathbb{C}^m$ be the homeomorphism given by

$$\alpha(z,w) = (\mathrm{Re}\ z,\ \mathrm{Im}\ z - F(w,w),\ w)$$

This restricts to a homeomorphism $\alpha : D(\Omega,F) \longrightarrow \mathbb{R}^n \times \Omega \times \mathbb{C}^m$ which shows that $D(\Omega,F)$ is a domain in $\mathbb{C}^{n+m}$.

Slightly heuristic considerations of $D(\Omega,F)$ for the case $m = 0$ allow us to make it a convention to regard $D(\Omega)$, the Siegel domain of the first kind associated to Ω, as the Siegel domain of the second kind associated to Ω and the zero form. The discussion for Siegel domains of the second kind holds in general for those of the first kind, and we shall not always mention this. So in the following we mean by a <u>Siegel domain</u> a Siegel domain of the first or second kind.

<u>Proposition 1.2</u>. Let $\Omega \subset \mathbb{R}^n$ be a convex open cone not containing any straight line and $F : \mathbb{C}^m \times \mathbb{C}^m \longrightarrow \mathbb{C}^n$ be an Ω-hermitian form. The Siegel domain of the second kind $D(\Omega,F)$ is isomorphic to a bounded domain in $\mathbb{C}^{n+m}$.

<u>Proof</u>. We have already seen that $D(\Omega,F)$ is a domain in $\mathbb{C}^{n+m}$. As before we may assume that $\Omega \subset \Omega'$, where $\Omega' = \{(x^1,\dots,x^n) \in \mathbb{R}^n \mid x^i > 0\}$. Then $D(\Omega,F) \subset D(\Omega',F)$. Hence to see that $D(\Omega,F)$ is isomorphic to a bounded domain we may assume that $\Omega = \Omega'$.

Now let

$$F(w,w') = (F^1(w,w'),\ F^2(w,w'),\dots,F^n(w,w'))$$

where

$$F^k(w,w') \in \mathbb{C}.$$

It is easy to see that the map $F^k : \mathbb{C}^m \times \mathbb{C}^m \longrightarrow \mathbb{C}$ is a non-negative definite hermitian form for each k, and $F^k(w,w) = 0$ for all k if and only if $w = 0$.

It is well known that any non-negative definite hermitian form B on $\mathbb{C}^m$ can be written as

$$B(w,w') = \sum_{j=1}^{m} B^j(w)\,\overline{B^j(w')}$$

where each B^j is a complex linear form on $\mathbb{C}^m$. Thus

$$F^k(w,w') = \sum_{j=1}^{m_k} L_k^j(w)\,\overline{L_k^j(w')},$$

L_k^j being complex linear forms on $\mathbb{C}^m$, and

$$F^k(w,w) = \sum_{j=1}^{m_k} |L_k^j(w)|^2$$

(We make convention that $m_k = 0$ if F^k is identically zero). Therefore F is determined completely by the forms $\{L_k^j\}_{k,j}$, and $L_k^j(w) = 0$ for all k and j if and only if $w = 0$.

This last observation means that the forms $\{L_k^j\}_{k,j}$ span the dual of $\mathbb{C}^m$ and hence there exist m linearly independent forms among these, say $L_1^1,\ldots,L_1^{p_1}, L_2^1,\ldots,L_n^{p_n}$ with $\sum_{k=1}^{n} p_k = m$. Define the map $\tilde{F}^k : \mathbb{C}^m \times \mathbb{C}^m \longrightarrow \mathbb{C}$ by putting

$$\tilde{F}^k(w,w') = \begin{cases} \sum_{j=1}^{p_k} L_k^j(w)\,\overline{L_k^j(w')} &, \text{ if } p_k > 0 \\ 0 &, \text{ if } p_k = 0. \end{cases}$$

Let

$$\tilde{F} = (\tilde{F}^1, \tilde{F}^2, \ldots \tilde{F}^n).$$

Obviously

$$0 \leqslant \tilde{F}^k(w,w) \leqslant F^k(w,w)$$

for all k, and $\tilde{F}^k(w,w) = 0$ for all k only if $w = 0$.

Therefore

$$\operatorname{Im} z^k - \tilde{F}^k(w,w) \geq \operatorname{Im} z^k - F^k(w,w) > 0,$$

for all k and for all $(z,w) \in D(\Omega,F)$. These observations show that $\tilde{F}$ is Ω-hermitian and we have $D(\Omega,F) \subset D(\Omega,\tilde{F})$ Now,

$$\begin{aligned} D(\Omega,\tilde{F}) &= \left\{(z,w) \in \mathbb{C}^n \times \mathbb{C}^m \mid \operatorname{Im} z - \tilde{F}(w,w) \in \Omega \right\} \\ &= \left\{(z,w) \in \mathbb{C}^n \times \mathbb{C}^m \mid \operatorname{Im} z^k - \tilde{F}^k(w,w) > 0 \quad \text{for all} \quad k \right\} \\ &= \left\{(z,w) \in \mathbb{C}^n \times \mathbb{C}^m \mid \operatorname{Im} z^k - \sum_{j=1}^{p_k} | L_k^j(w)|^2 > 0 \quad \text{for all} \quad k \right\} \\ &= \prod_k \left\{(z,w) \mid z \in \mathbb{C},\ w \in \mathbb{C}^{p_k},\ \operatorname{Im} z - \sum_{j=1}^{p_k} |L_k^j(w)|^2 > 0 \right\} \end{aligned}$$

If we chose a basis of $\mathbb{C}^m$ dual to $L_1^1,\ldots,L_1^{p_1},\ L_2^1,\ldots,L_2^{p_2},\ldots,L_n^1,\ldots,L_n^{p_n}$ then $\left\{ L_k^j(w) \mid 1 \leq j \leq p_k,\ k = 1,2\ldots n \right\}$ are coordinates of w. Therefore

$$\begin{aligned} D(\Omega,\tilde{F}) &= \prod_k \left\{(z,w) \mid z \in \mathbb{C},\ w \in \mathbb{C}^{p_k},\ \operatorname{Im} z - \sum_{i=1}^{p_k} |w^i|^2 > 0 \right\} \\ &= \prod_k \mathcal{E}^{p_k+1} \end{aligned}$$

We have observed that the elementary Siegel domain $\mathcal{E}^N$ is isomorphic to the unit open ball B^N in $\mathbb{C}^N$. Thus $D(\Omega,\tilde{F})$ is isomorphic to a bounded domain $\prod_k B^{p_k+1}$ in $\mathbb{C}^{n+m}$ and $D(\Omega,F)$ is isomorphic to a bounded domain.

(Q.E.D.)

§ 2. Affine automorphism groups of Siegel domains

We shall identify $\mathbb{C}^{n+m}$ with the product space $\mathbb{C}^n \times \mathbb{C}^m$ and denote elements of $\mathbb{C}^{n+m}$ by (z,w) where $z \in \mathbb{C}^n$, $w \in \mathbb{C}^m$. Let $AF(\mathbb{C}^{n+m})$ be the group of all affine automorphisms of $\mathbb{C}^{n+m}$. Then

$$AF(\mathbb{C}^{n+m}) = GL(\mathbb{C}^{n+m}).T(\mathbb{C}^{n+m}),$$

where $GL(\mathbb{C}^{n+m})$ is the group of invertible linear transformations of $\mathbb{C}^{n+m}$, $T(\mathbb{C}^{n+m})$ is the group of translations of $\mathbb{C}^{n+m}$ and '.' denotes the semidirect product. The group $AF(\mathbb{C}^{n+m})$ is a Lie group acting differentiably on $\mathbb{C}^{n+m}$.

Let $D = D(\Omega,F)$ be a Siegel domain in $\mathbb{C}^{n+m}$ associated to (Ω,F). Define

$$AF(D) = \left\{ g \in AF(\mathbb{C}^{n+m}) \;\middle|\; g(D) = D \right\}$$

Then $AF(D)$ is a closed subgroup of $AF(\mathbb{C}^{n+m})$. This follows from the fact that $g \in AF(\mathbb{C}^{n+m})$ belongs to $AF(D)$ if and only if it preserves the closure $\bar{D}$ of D and the boundary ∂D of D. Define

$$GL(D) = GL(\mathbb{C}^{n+m}) \cap AF(D).$$

We call elements of $AF(D)$ (resp. $GL(D)$) <u>affine</u> (resp. <u>linear</u>) <u>automorphisms</u> of D. Incidently it may be noted that $AF(D)$, $GL(D)$ are Lie groups, being closed subgroups of $AF(\mathbb{C}^{n+m})$, $GL(\mathbb{C}^{n+m})$ respectively.

To study the structure of $AF(D)$ it is convenient to consider the following automorphisms

i) For $a \in \mathbb{R}^n$ (contained in $\mathbb{C}^n$ canonically), define the map $p_a : \mathbb{C}^{n+m} \longrightarrow \mathbb{C}^{n+m}$ by

$$p_a(z,w) = (z + a, w)$$

ii) For $b \in \mathbb{C}^m$, define the map $q_b : \mathbb{C}^{n+m} \longrightarrow \mathbb{C}^{n+m}$ by

$$q_b(z,w) = (z + 2i\, F(w,b) + iF(b,b), w+b)$$

clearly $p_a (a \in \mathbb{R}^n)$ and $q_b (b \in \mathbb{C}^m)$ belong to $AF(\mathbb{C}^{n+m})$.

We make following observations.

Let $\Phi : \mathbb{C}^n \times \mathbb{C}^m \longrightarrow \mathbb{C}^n$ be the map defined by

$$\Phi(z,w) = \text{Im } z - F(w,w)$$

Then,

(2.1) $\Phi(p_a(z,w)) = \Phi(z,w) = \Phi(q_b(z,w))$

for any $a \in \mathbb{R}^n$, $b \in \mathbb{C}^m$. Therefore if $(z,w) \in D$, then $\Phi(z,w) \in \Omega$ and $\Phi(p_a(z,w)) = \Phi(q_b(z,w)) \in \Omega$, which means that $p_a(z,w)$, $q_b(z,w) \in D$. This proves that p_a, $q_b \in AF(D)$. We see immediately the <u>following</u> relations.

(2.2) $$\begin{cases} p_a p_{a'} = p_{a+a'} \\ p_a q_b = q_b p_a \\ q_b q_{b'} = q_{b+b'}\, p_{a^o} \quad \text{where } a^o = 2\text{Im } F(b,b') \end{cases}$$

for any $a, a' \in \mathbb{R}^n$ and $b, b' \in \mathbb{C}^m$.

Put

$$R = \left\{ p_a \mid a \in \mathbb{R}^n \right\}$$

$$P(D) = \left\{ p_a q_b \mid a \in \mathbb{R}^n,\ b \in \mathbb{C}^m \right\}.$$

Then (2.2) shows that R and $P(D)$ are subgroups of $AF(D)$. Also there is a canonical exact sequence of groups

(2.3) $$(0) \longrightarrow R \xrightarrow{\ i\ } P(D) \xrightarrow{\ \eta\ } \mathbb{C}^m \longrightarrow (0),$$

where i is the inclusion and η is given by $\eta(p_a q_b) = b$, which is a well defined homomorphism in view of (2.2). We see from this that $p(D)$ is a connected subgroup of $AF(D)$.

We note that every element of D can be transformed to an element of the form $(iy,0)$ where $y \in \Omega$, by an element of $P(D)$. In fact, if

$(z,w) \in D$, $a = \operatorname{Re} z$, then $q_{-w}\, p_{-a}\,(z,w) = (iy, 0)$ where y is the element of Ω given by $\operatorname{Im} z - F(w,w)$.

Lemma 2.1. Let $D = D(\Omega,F)$ be a Siegel domain. Then there exists a function f which is defined and holomorphic on the closure $\bar{D}$ of D (i.e. on an open neighbourhood of $\bar{D}$) and such that

$$|f(z,w)| \leq |f(0,0)| \quad \text{for all} \quad (z,w) \in \bar{D},$$

where the equality holds only if $(z,w) = (0,0)$.

Proof.

Case i) $m = 0$, and so D is a tube domain $D(\Omega) = \mathbb{R}^n + i\,\Omega$ in $\mathbb{C}^n$. We may assume that $\Omega \subset \Omega'$ the positive cone. Then the function f_o defined by

$$f_o(z) = \prod_{k=1}^{n} \frac{1}{z^k + i}$$

is holomorphic in a neighbourhood of $\overline{D(\Omega)}$ of the form $\mathbb{R}^n + iU_{\bar{\Omega}}$, where $U_{\bar{\Omega}}$ is a neighbourhood of $\bar{\Omega}$. Moreover, we see easily

$$|f_o(z)| \leq |f_o(0)| \qquad \text{for} \quad z \in \overline{D(\Omega)},$$

where the equality holds only if $z = 0$.

Case ii) General case, $D = D(\Omega,F)$: Let $U_{\bar{\Omega}}$ be a neighbourhood of $\bar{\Omega}$ as introduced in case i). We may further assume that $U_{\bar{\Omega}} + y \subset U_{\bar{\Omega}}$ for any $y \in \bar{\Omega}$. Now a neighbourhood $U_{\bar{D}}$ of $\bar{D}$ in $\mathbb{C}^{n+m}$ is defined as the set of all (z,w) such that

$$\operatorname{Im} z - F(w,w) \in U_{\bar{\Omega}}$$

Then, if $(z,w) \in U_{\bar{D}}$, $\operatorname{Im} z \in U_{\bar{\Omega}} + F(w,w) \subset U_{\bar{\Omega}}$. Thus we can define a holomorphic function f on $U_{\bar{D}}$ by putting

$$f(z,w) = f_o(z)$$

where f_o is the function obtained in case i). Then f is holomorphic on $U_{\bar{D}}$ and

$$|f(z,w)| = |f_o(z)| \leq |f_o(0)| = |f(0,0)|$$

for all $(z,w) \in \bar{D}$. If the equality holds for $(z,w) \in \bar{D}$ then $z = 0$ by the property of f_o, and hence $(o,w) \in \bar{D}$ so that $-F(w,w) \in \bar{\Omega}$. Since $\bar{\Omega}$ does not contain any straight line we get $F(w,w) = 0$ and therefore $w = 0$. Thus the function f satisfies the conditions of lemma 2.1.

(Q.E.D.)

As we remarked in § 1 the homeomorphism

$$\alpha : \mathbb{C}^n \times \mathbb{C}^m \longrightarrow \mathbb{R}^n \times \mathbb{R}^n \times \mathbb{C}^m$$

defined by

$$\alpha(z,w) = (\mathrm{Re}\ z, \Phi(z,w), w)$$

induces a bijection of $D = D(\Omega,F)$ with the open set $\mathbb{R}^n \times \Omega \times \mathbb{C}^m$. Therefore the closure $\bar{D}$ of D corresponds under α to the set $\mathbb{R}^n \times \bar{\Omega} \times \mathbb{C}^m$. This proves the relations,

$$\bar{D} = \left\{(z,w) \in \mathbb{C}^{n+m} \,|\, \Phi(z,w) \in \bar{\Omega}\right\}$$

and

$$\partial D = \left\{(z,w) \in \mathbb{C}^{n+m} \,|\, \Phi(z,w) \in \partial\Omega\right\}$$

where ∂D and $\partial\Omega$ denote boundaries of D and Ω respectively. In particular, one sees that the origin $(0,0)$ belongs to $\partial D \subset \bar{D}$.

Defintion. The set

$$S = \left\{(z,w) \in \mathbb{C}^{n+m} \,|\, \Phi(z,w) = 0\right\}$$

is called the Šilov boundary of the Siegel domain $D(\Omega,F)$.

By the above remark, S is contained in ∂D.

Lemma 2.2. Let $D = D(\Omega,F)$ be a Siegel domain and S be its Šilov boundary.

Then

$$S = \{g(0,0) \mid g \in P(D)\} = \{g(0,0) \mid g \in AF(D)\}$$

Proof. By (2.1), we have

$$\Phi(0,0) = \Phi(g(0,0)) \quad \text{for any} \quad g \in P(D).$$

It follows that the set $\{g(0,0) \mid g \in P(D)\}$ is contained in S. Next consider $(z,w) \in S$. Then $\Phi(z,w) = 0$ and so $\operatorname{Im} z = F(w,w)$. Put $a = \operatorname{Re} z$. Then

$$\begin{aligned} q_{-w}\, p_{-a}\, (z,w) &= q_{-w}(i \operatorname{Im} z, w) \\ &= (i \operatorname{Im} z + 2i\, F(w,-w) + iF(-w,-w),\ w-w) \\ &= (0,0) \end{aligned}$$

and so $(z,w) = p_{-a}^{-1}\, q_{-w}^{-1}\, (0,0)$ belongs to the set $\{g(0,0) \mid g \in P(D)\}$.

We next show that the set $\{g(0,0) \mid g \in AF(D)\}$ is contained in set $\{g(0,0) \mid g \in P(D)\}$. Let $g \in AF(D)$ and $g(0,0) = (z_o, w_o)$. Since $(0,0) \in \bar{D}$, $(z_o, w_o) \in \bar{D}$. Take $g' = q_{-w_o}\, p_{-a_o}\, g$ where $a_o = \operatorname{Re} z_o$. Then

$$g'(0,0) = q_{-w_o} p_{-a_o} g(0,0) = q_{-w_o} p_{-a_o} (z_o, w_o) = (iy_o, 0)$$

with $y_o = \operatorname{Im} z_o - F(w_o, w_o) \in \bar{\Omega}$ and $(iy_o, 0) \in \bar{D}$.

We claim that $y_o = 0$. For this, consider the function f in Lemma 2.1 for $D(\Omega, F)$. Put

$$f'(z,w) = f(g'^{-1}(z,w)) \quad \text{for} \quad (z,w) \in \bar{D}.$$

Then f' is holomorphic on $\bar{D}$. Moreover

$$\begin{aligned} |f'(z,w)| &= |f(g'^{-1}(z,w))| \\ &\le |f(0,0)| \\ &= |f(g'^{-1}(iy_o, 0))| \\ &= |f'(iy_o, 0)| \end{aligned}$$

and $|f'(z,w)| = |f'(iy_o,0)|$ only if $(z,w) = (iy_o,0)$. Now for λ varying over the upper half plane, let $\phi(\lambda) = f'(\lambda y_o,0)$. Suppose $y_o \neq 0$. Then by the properties of f', ϕ is holomorphic, $|\phi(\lambda)| \leq |\phi(i)|$ and equality holds only if $\lambda = i$. This contradicts the maximal principle of holomorphic functions in one complex variable. Therefore $y_o = 0$ and $g'(0,0) = (0,0)$. This shows that $g(0,0) = p^{-1}_{-a_o} q^{-1}_{-w_o} (0,0)$ and hence $g(0,0) \in P(D)(0,0)$ and we have

$$\left\{ g(0,0) \mid g \in AF(D) \right\} \subset \left\{ g(0,0) \mid g \in P(D) \right\}$$

The converse inclusion being trivial, we have proved the lemma 2.2.

(Q.E.D.)

<u>Proposition 2.1</u>. Let D be a Siegel domain. Then

$$AF(D) = GL(D).P(D) \qquad \text{(semi-direct)}$$

where $P(D)$ is a normal subgroup of $AF(D)$.

<u>Proof</u>. Let $g \in AF(D)$. Then there exists $g' \in P(D)$ such that $g(0,0) = g'(0,0)$ by lemma 2.2. Therefore $g'^{-1}g$ fixes zero which shows that $g'^{-1}g \in GL(\mathbb{C}^{n+m})$. Hence $g'^{-1}g \in GL(D)$. Therefore $g = g'(g'^{-1}g) \in P(D)GL(D)$.

We now show that $GL(D) \cap P(D) = (e)$. For this, let $p_a q_b \in GL(D)$ then $p_a q_b(0,0) = (0,0)$, that is $(a + iF(b,b),b) = (0,0)$ which means $a = b = 0$ or $p_a q_b = e$ (the identity of $AF(D)$) and hence

$$GL(D) \cap P(D) = (e).$$

We see easily that $P(D)$ is a normal subgroup of $AF(D)$.

(Q.E.D.)

<u>Corollary.</u> The connected component G_a of the identity in $AF(D)$ is given by

$$G_a = GL(D)^o . P(D) \qquad \text{(semi-direct)}$$

where $GL(D)^o$ is the connected component of the identity in $GL(D)$.

Proof. As followed from (2.3), $P(D)$ is a connected subgroup and hence is contained in G_a. Then by Proposition 2.1

$$G_a = H . P(D)$$

for some subgroup H of $GL(D)$. Since $H.P(D)$ is connected it follows that $H = GL(D)^o$.

(Q.E.D.)

Let $\Omega \subset \mathbb{R}^n$ be an open convex cone not containing a straight line. Define

$$\operatorname{Aut} \Omega = \left\{ A \in GL(\mathbb{R}^n) \mid A(\Omega) = \Omega \right\}$$

The canonical inclusion $\mathbb{R}^n \subset \mathbb{C}^n$ induces the inclusion of $GL(\mathbb{R}^n)$ in $GL(\mathbb{C}^n)$. Since we identify $\mathbb{C}^{n+m}$ with $\left\{(z,w) \mid z \in \mathbb{C}^n, w \in \mathbb{C}^m\right\}$ we may consider $GL(\mathbb{C}^n) \times GL(\mathbb{C}^m)$ as a subgroup of $GL(\mathbb{C}^{n+m})$ by putting $(A,B)(z,w) = (Az,Bw)$ for $(A,B) \in GL(\mathbb{C}^n) \times GL(\mathbb{C}^m)$. We shall also identify the linear transformation in $\mathbb{C}^n$ or $\mathbb{C}^{n+m}$ with the matrix representing the transformation with respect to canonical coordinates.

Proposition 2.2. Let $D = D(\Omega, F)$ be a Siegel domain in $\mathbb{C}^{n+m}$. Then $GL(D)$ consists of all elements $(A,B) \in GL(\mathbb{C}^n) \times GL(\mathbb{C}^m)$ satisfying the two conditions:

i) $A \in \operatorname{Aut} \Omega$

ii) $A\, F(w,w) = F(Bw,Bw)$ for all $w \in \mathbb{C}^m$

Proof. If (A,B) satisfies i) and ii) then for $(z,w) \in D$ we have

$$\begin{aligned} \Phi(Az,Bw) &= \operatorname{Im} Az - F(Bw,Bw) \\ &= A \operatorname{Im} z - AF(w,w) \\ &= A(\operatorname{Im} z - F(w,w)) \end{aligned}$$

$$= A(\Phi(z,w))$$

Since $\Phi(z,w) \in \Omega$ and $A \in \mathrm{Aut}\,\Omega$, $A(\Phi(z,w)) \in \Omega$. This shows that $(Az,Bw) \in D$. Since $(A,B) \in GL(\mathbb{C}^{n+m})$ it follows that $(A,B) \in GL(D)$.

Conversely let $g \in GL(D) \subset GL(\mathbb{C}^{n+m})$. Then we may write

$$g(z,w) = (Az + C_1 w,\ C_2 z + Bw)$$

where $A \in GL(\mathbb{C}^n)$, $B \in GL(\mathbb{C}^m)$, $C_1 \in \mathrm{Hom}\,(\mathbb{C}^m, \mathbb{C}^n)$ and $C_2 \in \mathrm{Hom}\,(\mathbb{C}^n, \mathbb{C}^m)$.

("Hom" denotes the set of all linear homomorphisms).

i) A is real: By lemma 2.2 $S = AF(D)(0,0)$ and $AF(D)$ preserves S. Let $(x,0) \in \mathbb{R}^n \times \{0\} \subset S$. Then

$$g(x,0) = (Ax, C_2 x) \in S$$

which implies that $\mathrm{Im}\, Ax = F(C_2 x, C_2 x) \in \bar{\Omega}$. Replacing x by $-x$ in this we get that $-\mathrm{Im}\, Ax \in \bar{\Omega}$. Since Ω does not contain any straight line, it follows that $\mathrm{Im}\, Ax = 0$. Since this holds for any $x \in \mathbb{R}^n$, A is a real matrix.

ii) $C_2 = 0$: Since A is real, $F(C_2 x, C_2 x) = \mathrm{Im}\, Ax = 0$ for all $x \in \mathbb{R}^n$. Therefore $C_2 x = 0$ for all $x \in \mathbb{R}^n$. Thus $C_2 = 0$.

iii) $A \in \mathrm{Aut}\,\Omega$: If $y \in \Omega$ then $(iy, 0) \in D$. Since $g \in GL(D)$, $g(iy,0) \in D$ and $(iAy,0) \in D$. Thus $\mathrm{Im}\,(iAy) \in \Omega$. As we have seen that A is real, it follows $Ay \in \Omega$, which proves $A \in \mathrm{Aut}\,\Omega$

iv) $AF(w,w) = F(Bw,Bw)$ and $C_1 = 0$; Let $w \in \mathbb{C}^m$, and $y = F(w,w) \in \bar{\Omega}$. Then $(iy,w) \in S$. By lemma 2.2 $g(iy,w) = (iAy + C_1 w, Bw) \in S$. Thus

$$\mathrm{Im}\; iAy + \mathrm{Im}\; C_1 w - F(Bw,Bw) = 0$$

Replacing w in this by $e^{it} w$ $(t \in \mathbb{R})$, we get

$$\mathrm{Im}\; C_1 e^{it} w = F(Bw,Bw) - Ay \in \mathbb{R}^n$$

This being true for all $t \in \mathbb{R}$, it follows that

$$\operatorname{Im} C_1 e^{it} w = F(Bw,Bw) - Ay = 0.$$

The last equality means that

$$AF(w,w) = F(Bw,Bw)$$

Also $\operatorname{Im} C_1 e^{it} w = \operatorname{Im} e^{it} C_1 w = 0$ for all $t \in \mathbb{R}$. Therefore $C_1 w = 0$. Since this holds for all $w \in \mathbb{C}^m$, we get $C_1 = 0$.

These four claims together prove the Proposition 2.2.

(Q.E.D.)

The affine group $AF(\mathbb{C}^{n+m})$ is isomorphic canonically to the subgroup

$$\left\{ \begin{pmatrix} A & \begin{matrix} z \\ w \end{matrix} \\ 0 \cdots 0 & 1 \end{pmatrix} \;\middle|\; A \in GL(\mathbb{C}^{n+m}), \text{ and } z \in \mathbb{C}^n, w \in \mathbb{C}^m \right\}$$

of $GL(\mathbb{C}^{n+m+1})$, an element of $\mathbb{C}^n$ (resp. $\mathbb{C}^m$) being represented by an $(n,1)$ (resp. $(m,1)$) matrix. The Lie algebra $\mathfrak{a}$ of $AF(\mathbb{C}^{n+m})$ is given by

$$\mathfrak{a} = \left\{ \begin{pmatrix} X & \begin{matrix} z \\ w \end{matrix} \\ 0 \cdots 0 & 0 \end{pmatrix} \;\middle|\; X \in M_{n+m}(\mathbb{C}),\ z \in \mathbb{C}^n,\ w \in \mathbb{C}^m \right\}$$

where $M_{n+m}(\mathbb{C})$ denotes that set of all complex matrices of degree $n+m$.

Let $D = D(\Omega,F)$ be a Siegel domain contained in $\mathbb{C}^{n+m}$. The affine automorphism group $AF(D)$ of D is a closed subgroup of $AF(\mathbb{C}^{n+m})$, hence a Lie subgroup. Let $\mathfrak{a}(D)$ be the Lie algebra of $AF(D)$. We proceed to study the structure of $\mathfrak{a}(D)$.

In the group $AF(D)$ consider following one-parameter subgroups

i) $\{p_{ta}\}_{t \in \mathbb{R}}$ $(a \in \mathbb{R}^n)$, where $p_{ta}(z,w) = (z + ta, w)$

ii) $\{q_{tb}\}_{t\in\mathbb{R}}$ $(b \in \mathbb{C}^m)$, where $q_{tb}(z,w) = (z + 2i\,F(w,tb) + iF(tb,tb),\; w + tb)$

iii) $\{c_t\}_{t\in\mathbb{R}}$, where $c_t(z,w) = (e^t z, e^{\frac{1}{2}t} w)$

By proposition 2.2 we see that $c_t \in GL(D)$, for all $t \in \mathbb{R}$.

Define

$\mathfrak{A}_{-1}$ = set of elements of $\mathfrak{A}(D)$ defined by one-parameter subgroups $\{p_{ta}\}_{t\in\mathbb{R}}$, a varying over $\mathbb{R}^n$.

$\mathfrak{A}_{-\frac{1}{2}}$ = set of elements of $\mathfrak{A}(D)$ defined by one parameter subgroups $\{q_{tb}\}_{t\in\mathbb{R}}$, b varying over $\mathbb{C}^m$.

and let $E \in \mathfrak{A}(D)$ be the element defined by $\{c_t\}_{t\in\mathbb{R}}$.

We now characterise $\mathfrak{A}_{-1}$ and $\mathfrak{A}_{-\frac{1}{2}}$ as subsets of $\mathfrak{A}(D)$, represented as matrix Lie algebra.

i) Since every p_a $(a \in \mathbb{R}^n)$ is represented by $\begin{pmatrix} 1_n & 0 & a \\ 0 & 1_m & 0 \\ 0 & 0 & 1 \end{pmatrix}$

we see $\mathfrak{A}_{-1} = \left\{ \begin{pmatrix} 0 & a \\ & 0 \\ 0\;0 & 0 \end{pmatrix} \;\middle|\; a \in \mathbb{R}^n \right\}$

ii) Similarly we see

$$\mathfrak{A}_{-\frac{1}{2}} = \left\{ \begin{matrix} n \\ m \\ \\ \end{matrix}\begin{pmatrix} 0 & Y_b & 0 \\ 0 & 0 & b \\ 0 & & 0 \end{pmatrix} \quad b \in \mathbb{C}^m \right\}$$

where Y_b is the $n \times m$ matrix representing the transformation defined by $Y_b(w) = 2i\,F(w,b)$.

It follows from this characterisation that $\mathfrak{A}_{-1}$ and $\mathfrak{A}_{-\frac{1}{2}}$ are subspaces of $\mathfrak{A}(D)$. It is easy to verify that the one parameter group

$\{c_t\}_{t\in\mathbb{R}}$ defines the element $E = \begin{pmatrix} 1_n & 0 & 0 \\ 0 & \frac{1}{2}1_m & 0 \\ 0 & 0 & 0 \end{pmatrix}$ where 1_n (resp. 1_m) denotes the unit matrix of degree n (resp. m). Thus we have

$$(2.4) \quad [E,X] = EX - XE = \begin{cases} X & \text{if } X \in \mathfrak{a}_{-1} \\ \frac{1}{2}X & \text{if } X \in \mathfrak{a}_{-\frac{1}{2}} \\ 0 & \text{if } X \in \mathfrak{a}_0 \end{cases}$$

where $\mathfrak{a}_0$ is the Lie subalgebra corresponding to the closed subgroup $GL(D)$ of $AF(D)$. (Last equality is proved using Proposition 2.2 from where one verifies that an element of $\mathfrak{a}_0$ is of the form $\begin{pmatrix} A & 0 & 0 \\ 0 & B & 0 \\ 0 & 0 & 0 \end{pmatrix}$ for some matrices A and B.).

This analysis shows that $\mathfrak{a}_{-1}$, $\mathfrak{a}_{-\frac{1}{2}}$, $\mathfrak{a}_0$ are respectively contained in the eigenspaces of $-\mathrm{ad}\, E$ with eigenvalues -1, $-\frac{1}{2}$, 0.

We now prove

<u>Proposition 2.3</u>. Let $D = D(\Omega,F)$ be a Siegel domain and let $\mathfrak{a}(D)$ be the Lie algebra of the group of all affine automorphisms of D. Then $\mathfrak{a}(D)$ decomposes to a direct sum of real vector subspaces

$$\mathfrak{a}(D) = \mathfrak{a}_{-1} + \mathfrak{a}_{-\frac{1}{2}} + \mathfrak{a}_0$$

Moreover $\mathfrak{a}_{-1}$, $\mathfrak{a}_{-\frac{1}{2}}$, $\mathfrak{a}_0$ are eigenspaces of $-\mathrm{ad}\, E$ with eigenvalues -1, $-\frac{1}{2}$ and 0 respectively.

<u>Proof</u>. Since $AF(D) = GL(D).P(D)$ (semidirect product) by Proposition 2.1, it follows that $\mathfrak{a}(D) = \mathfrak{a}_0$ + Lie algebra of $P(D)$ (direct sum as real vector space). To compute the Lie algebra of $P(D)$ consider the exact sequence (2.3)

$$(0) \longrightarrow R \longrightarrow P(D) \longrightarrow \mathbb{C}^m \longrightarrow (0)$$

It follows from this that the real dimension of $P(D)$ is $2m+n$. Now $\mathfrak{A}_{-1}$ and $\mathfrak{A}_{-\frac{1}{2}}$ are contained in the Lie algebra of $P(D)$ (as corresponding one-parameter groups are in $P(D)$) and are of dimension n and $2m$ respectively. Moreover $\mathfrak{A}_{-1} \cap \mathfrak{A}_{-\frac{1}{2}} = (0)$; since $\mathfrak{A}_{-1}$ and $\mathfrak{A}_{-\frac{1}{2}}$ are contained in the eigen spaces of $-\mathrm{ad}\, E$ with distinct eigenvalues. Thus the dimension of $\mathfrak{A}_{-1} + \mathfrak{A}_{-\frac{1}{2}}$ is $2m + n$ and therefore the Lie algebra of $P(D)$ coincides with $\mathfrak{A}_{-1} + \mathfrak{A}_{-\frac{1}{2}}$. Thus we have proved $\mathfrak{A}(D) = \mathfrak{A}_{-1} + \mathfrak{A}_{-\frac{1}{2}} + \mathfrak{A}_0$.

Second part of the proposition follows from the first part and (2.4).

(Q.E.D.)

Lastly we prove the following

<u>Proposition 2.4</u>. Let $D = D(\Omega, F)$ be a Siegel domain. Let g be a holomorphic automorphism of D which commutes with every element of the form p_a, q_b, c_t for any $a \in \mathbb{R}^n$, $b \in \mathbb{C}^m$, $t \in \mathbb{R}$. Then g is the identity automorphism.

<u>Proof</u>. We introduce some notation for convenience. Write (z', w') for $g(z,w)$ and put

$$z' = (g^1(z,w), g^2(z,w) \dots g^n(z,w)),$$

$$w' = (g^{n+1}(z,w), g^{n+2}(z,w) \dots g^{n+m}(z,w))$$

Also we use k, ℓ for indexing integers from 1 to n and α, β for indexing integers from $n+1$ to $n+m$, without mentioning it every time. Now, by assumption

$$p_{ta}\, g(z,w) = g p_{ta}\, (z,w)$$

for any $a \in \mathbb{R}^n$ and $t \in \mathbb{R}$, namely

$$(z' + ta, w') = ((z + ta)', w').$$

Comparing the components

$$\text{(i)} \qquad g^k(z + ta, w) = g^k(z,w) + ta^k.$$

$$\text{(ii)} \qquad g^\alpha(z + ta, w) = g^\alpha(z,w)$$

Differentiating (i) with respect to t we get

$$\sum_{\ell=1}^{n} \frac{\partial g^k}{\partial z^\ell} a^\ell = a^k \quad \text{for} \quad a = (a^1, \dots, a^n) \in \mathbb{R}^n$$

whence we get

$$(*) \qquad \frac{\partial g^k}{\partial z^\ell} = \delta_k^\ell .$$

Differentiating (ii) with respect to t

$$\sum_{\ell=1}^{n} \frac{\partial g^\alpha}{\partial z^\ell} a^\ell = 0 \quad \text{for} \quad a = (a^1, \dots, a^n) \in \mathbb{R}^n$$

Therefore $\frac{\partial g^\alpha}{\partial z^\ell} = 0$ for all ℓ and g^α is a function of w alone,

$g^\alpha(z,w) = g^\alpha(w)$ say.

Secondly, again by assumption

$$q_{tb}\, g(z,w) = g q_{tb}(z,w)$$

for any $b \in \mathbb{C}^m$ and $t \in \mathbb{R}$, namely

$$(z' + 2i\,F(w',tb) + iF(tb,\, tb),\ w' + tb)$$

$$= ((z + 2i\,F(w,tb) + iF(tb,\, tb))',\ (w + tb)')$$

Therefore

$$g^\alpha(z,w) + tb^{\alpha - n} = g^\alpha(\dots, w + tb)$$

$$= g^\alpha(w + tb)$$

Differentiating with respect to t, and writing z^β for $w^{\beta - n}$, we get

$$b^{\alpha - n} = \sum_\beta \frac{\partial g^\alpha}{\partial z^\beta} b^{\beta - n} \quad \text{for all} \quad b \in \mathbb{C}^m$$

which implies that $\frac{\partial g^\alpha}{\partial z^\beta} = \delta^\beta_\alpha$. Therefore $g^\alpha(w) = z^\alpha + d^\alpha$ where d^α are constants.

Comparing k components and differentiating with respect to t at $t = 0$, we get by $(*)$

$$\frac{\partial g^k}{\partial z^\alpha} = 0$$

and hence $g^k(z,w) = g^k(z)$ say. Thus

$$g(z,w) = (g^1(z), \ldots g^n(z), z^{n+1} + d^{n+1}, \ldots, z^{n+m} + d^{n+m}).$$

Lastly

$$gc_t(z,w) = c_t g(z,w)$$

i.e.

$$g^k(e^t z) = e^t(g^k(z)) \quad \text{and} \quad e^{\frac{1}{2}t} z^\alpha + d^\alpha = e^{\frac{1}{2}t}(z^\alpha + d^\alpha).$$

Differentiating with respect to t at $t = 0$ we get

$$g^k(z) = z^k \quad \text{and} \quad d^\alpha = 0$$

for all k and α respectively. Thus $g(z,w) = (z^1, \ldots z^n, z^{n+1}, \ldots, z^{n+m}) = (z,w)$ i.e. g is the identity automorphism.

(Q.E.D.)

Corollary. Any subgroup of the group of automorphisms $G(D)$ of D containing the subgroup $P(D)$ and the one-parameter subgroup $\{c_t\}_{t \in \mathbb{R}}$ has trivial centre. In particular, the following groups have trivial centres: $G(D)$, $AF(D)$ and their connected components of the identity.

Proof. This follows immediately from the Proposition.

§ 3. Affinely homogeneous Siegel domains

Definition. A Siegel domain $D = D(\Omega, F)$ is said to be affinely homogeneous if the group $AF(D)$ acts transitively on D.

In this case the connected component of the identity of $AF(D)$ acts transitively on D.

Using the characterization of $GL(D)$ in Proposition 2.2 we can define a homomorphism μ of $GL(D)$ into $\mathrm{Aut}\,\Omega$ by

$$\mu(A,B) = A$$

<u>Proposition 3.1</u>. A Siegel domain $D = D(\Omega,F)$ is affinely homogeneous if and only if $\mu(GL(D))$ acts transitively on Ω.

<u>Proof</u>. Suppose first that $\mu(GL(D))$ acts transitively on Ω. Let (z,w) and (z',w') be arbitrary points of D. Then as remarked in § 2 there exist elements g_1, g_2 of $P(D)$ such that

$$g_1(z,w) = (iy,0) \quad \text{and} \quad g_2(z',w') = (iy',0)$$

for some $y, y' \in \Omega$.

By assumption there exists an element g in $GL(D)$ such that $\mu(g)(y) = y'$. Hence $g(iy,0) = (iy',0)$, so that

$$g_2^{-1} g\, g_1(z,w) = (z',w')$$

Thus $AF(D)$ acts transitively on D, that is, D is affinely homogeneous. Conversely suppose that D is affinely homogeneous. Let $y, y' \in \Omega$. There exists an element g in $AF(D)$ such that $g(iy,0) = (iy',0)$. By Proposition 2.1 $AF(D) = GL(D).P(D)$, so that $g = g_1 . g_2$ where $g_1 \in GL(D)$ and $g_2 \in P(D)$. Let $g_2 = p_a q_b$ $(a \in \mathbb{R}^n,\ b \in \mathbb{C}^m)$. Thus we have

$$g_1\, p_a q_b(iy,0) = (iy',0)$$

i.e.

$$g_1(iy + a + iF(b,b),b) = (iy',0)$$

Since g_1 belongs to $GL(D)$, it follows from Proposition 2.2 that $b = 0$ and

$a = 0$. Thus $g = g_1$, and $g_1(iy,0) = (iy',0)$, which shows that $\mu(g_1)(y) = y'$. Therefore $\mu(GL(D))$ acts transitively on Ω.

(Q.E.D.)

Corollary. A Siegel domain of the first kind $D = D(\Omega)$ is affinely homogeneous if and only if $\mathrm{Aut}\,\Omega$ acts transitively on Ω (i.e. Ω is homogeneous).

Proof. This follows immediately from Proposition, since $GL(D) = \mathrm{Aut}\,\Omega$ in this case.

Proposition 3.2. Let D be an affinely homogeneous Siegel domain, and G_a be the connected component of the identity of $AF(D)$. Then G_a coincides with the connected component of the identity of the normalizer $N(G_a)$ of G_a in $AF(\mathbb{C}^{n+m})$.

Proof. Let us denote by $N(G_a)^o$ the connected component of the identity in $N(G_a)$. We have only to prove that $N(G_a)^o \subset G_a$.

Since $AF(\mathbb{C}^{n+m})$ acts continuously on $\mathbb{C}^{n+m}$ so does $N(G_a)^o$. Therefore for x_o belonging to D, there exists a neighbourhood V of identity in $N(G_a)^o$ such that if $g \in V$ then $gx_o \in D$. Since $g \in N(G_a)^o$

$$G_a g x_o = g(g^{-1}G_a g)x_o = gG_a x_o$$

Since G_a acts transitively on D, $G_a g x_o = D$ and $G_a x_o = D$. Therefore

$$D = g(D)$$

Since g belongs to $AF(\mathbb{C}^{n+m})$ it follows that $g \in AF(D)$. Hence V is contained in $AF(D)$. Since $N(G_a)^o$ is connected, it is generated by V. Hence $N(G_a)^o \subset AF(D)$. The group G_a being the connected component of the identity of $AF(D)$, we get

$$N(G_a)^o \subset G_a$$

(Q.E.D.)

Remark 1. One can show from this Proposition that G_a coincides with the connected component of the identity of a real algebraic subgroup of $AF(\mathbb{C}^{n+m})$.

Proposition 3.3. Let D be an affinely homogeneous Siegel domain. Then the isotropy group of G_a at a point of D is a maximal compact subgroup of G_a.

Proof. It is sufficient to show the proposition for a point $(iy_0,0) \in D$ for some $y_0 \in \Omega$, since the isotropy group at any other point of D is a conjugate to the isotropy group at $(iy_0,0)$.

Let K be the isotropy group of G_a at $(iy_0,0)$. It is known that the isotropy group of G(D), the group of holomorphic automorphisms of D, at a point of D, is compact. Since G_a is a closed subgroup of G(D) and K is the intersection of G_a with the isotropy group of G(D) at $(iy_0,0)$, it follows that K is a compact subgroup.

We next show that K is a maximal compact subgroup. Let μ be the homomorphism of GL(D) into Aut Ω defined in Proposition 3.1. It is obvious that the kernel of μ is in K. Now by the corollary to Proposition 2.1 we have,

$$G_a = GL(D)^o.P(D) \quad \text{(semidirect product)}$$

It follows by the same argument as in the proof of Proposition 3.1 that every element of K belongs to $GL(D)^o$. Thus $K \subset GL(D)^o$.

Now put $K' = \mu(K)$. Then K' is the isotropy of $G' = \mu(GL(D)^o)$ at $y_0 \in \Omega$. By Proposition 3.1, D is affinely homogeneous if and only if $G' = \mu(GL(D)^o)$ acts transitively on Ω. Thus G'/K' is homeomorphic to Ω. Since Ω is simply connected it follows that K' is a connected subgroup in Aut Ω.

Let now L be any compact subgroup of G_a containing K. Then $L' = \mu(L)$ contains K' and is a compact subgroup of Aut Ω. Thus $L'y_0 = \{\ell' y_0 | \ell' \in L'\}$ is a compact subset of Ω. Let $[L'y_0]$ be the convex closure of $L'y_o$ in $\mathbb{R}^n$. Since Ω is convex $[L'y_o] \subset \Omega$. On the other hand it is known that the convex closure of a compact set in $\mathbb{R}^n$ is again a compact set. (See Eggleston, Convexity, Cambridge University Press, 1958). Let ν be the Haar measure on the compact group L' such that $\nu(L') = 1$. Let

$$y_1 = \int_{L'} \ell'(y_o)\, d\nu$$

Then, since $[L'y_o]$ is a closed set, we have

$$y_1 \in [L'y_o] \subset \Omega .$$

By the invarience of the Haar measure it follows that y_1 is fixed under the action of the group L'. Now, since G' acts transitively on Ω, there exists $g' \in G'$ such that $g'y_o = y_1$. Then we get

$$g'^{-1} L' g' \subset K' \quad \text{and so}$$

$$K' \subset L' \subset g' K' g'^{-1}$$

Since K' and $g' K' g'^{-1}$ are connected Lie subgroups of same dimension in Aut Ω, it follows that $K' = L' = g'K'g'^{-1}$. So $\mu(K) = \mu(L)$. Since the kernel of μ is contained in K, we get $K = L$, which proves that K is a maximal compact subgroup of G_a.

Remark 2. A theorem of Vinberg (Soviet Math Doklady 2 (1961) 1416) states - Let G be the connected component of the identity of a real algebraic subgroup of $GL(n,\mathbb{R})$ and K be a maximal compact subgroup of G. Then there exists a triangular subgroup (i.e. a group of upper triangular matrices) T such that $G = T.K$ and $T \cap K = (e)$.

Using this theorem and the results announced in Remark 1, we see from the above proposition that there exists a subgroup T of $AF(D)$ which acts simply transitively on D.

Examples of affinely homogeneous Siegel domains.

We have already proved that $H^+(l, \mathbb{R})$, the set of positive definite real symmetric matrices, is an open convex cone not containing any straight line in $H(l, \mathbb{R})$, the space of real symmetric matrices of degree l. Let $M(l,q,\mathbb{C})$ be the complex vector space of all $l \times q$ complex matrices. Take for $\mathbb{C}^m$ the space $M(l,q,\mathbb{C})$ so that $m = lq$. Define the map $F : \mathbb{C}^m \times \mathbb{C}^m \longrightarrow H(l,\mathbb{C})$ by

$$F(w,v) = \tfrac{1}{2}\,(w\,{}^t\bar{v} + \bar{v}\,{}^t w)$$

Then

$$F(w,w) = \tfrac{1}{2}\,(w\,{}^t\bar{w} + \bar{w}\,{}^t w)$$

and $F(w,w)$ being non-negative definite real matrix belongs to the closure of $H^+(l,\mathbb{R})$ in $H(l,\mathbb{R})$. Thus F is an $H^+(l,\mathbb{R})$-hermitian form on $\mathbb{C}^m = M(l,q,\mathbb{C})$.

Put $D = D(\Omega,F)$ with Ω and F as defined above. For each element $a \in GL(l,\mathbb{R})$ we can define an automorphism A_a of $\Omega = H^+(l,\mathbb{R})$ by

$$A_a(y) = a y\,{}^t a \qquad (y \in \Omega)$$

Again for each element $a \in GL(l,\mathbb{R})$ define the linear transformation B_a of $M(l,q,\mathbb{C})$ by $B_a(w) = aw$. Then

$$\begin{aligned} A_a F(w,w) &= a\left\{\tfrac{1}{2}\,(w\,{}^t\bar{w} + \bar{w}\,{}^t w)\right\}{}^t a \\ &= \tfrac{1}{2}\left\{a w\,{}^t\bar{w}\,{}^t a + a\bar{w}\,{}^t w\,{}^t a\right\} \\ &= F(B_a w, B_a w) \end{aligned}$$

which shows that $(A_a,B_a) \in GL(D)$. Hence $\left\{(A_a,B_a) \mid a \in GL(l,\mathbb{R})\right\} \subset GL(D)$.

Because any positive definite real symmetric matrix is of the form $a\,{}^t a$ with $a \in GL(l,\mathbb{R})$, it follows that the set $\left\{ A_a \mid a \in GL(l,\mathbb{R})\right\}$ acts transitively on $H^+(l,\mathbb{R})$. Therefore, $\mu(GL(D))$ acts transitively on $H^+(l,\mathbb{R})$ and by Proposition 3.1, D is affinely homogeneous.

From this example we can construct more examples as follows. For $\mathbb{C}^m$ take a subspace of $M(\ell,q,\mathbb{C})$ for which there exists a subgroup H of $GL(\ell,\mathbb{R})$ such that $\mathbb{C}^m$ is invariant under $\{B_a | a \in H\}$ and that $\{A_a | a \in H\}$ acts transitively on $\Omega = H^+(\ell,\mathbb{R})$. With the above Ω and F restricted to $\mathbb{C}^m \times \mathbb{C}^m$, we get a Siegel domain $D = D(\Omega, F)$. Again $(A_a, B_a) \in GL(D)$ for all $a \in H$ and since $\{A_a \backslash a \in H\}$ acts transitively on Ω, D is affinely homogeneous by Proposition 3.1. In particular, let H be the subgroup of all upper triangular matrices in $GL(\ell,\mathbb{R})$. Since $GL(\ell,\mathbb{R})$ is the product of H and the orthogonal group $O(\ell)$ it follows that $\{A_a | a \in H\}$ acts transitively on Ω. Let $\mathbb{C}^m$ be the subspace of $M(\ell,q,\mathbb{C})$ consisting of matrices of the form

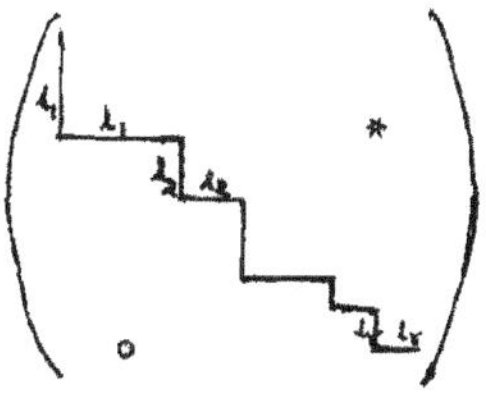

where $\ell_1, \ldots, \ell_r$ are positive integers such that $\sum_{j=1}^{r} \ell_j = \text{Min } \ell, q$.

It is easy to verify that $\mathbb{C}^m$ is invariant under $\{B_a | a \in H\}$. Since $\{A_a | a \in H\}$ acts transitively on $\Omega = H^+(\ell,\mathbb{R})$, the Siegel domain $D = D(\Omega, F/\mathbb{C}^m)$ is affinely homogeneous.

In the special case where $q = 2$ and $\ell = \ell_1 = 1$ this Siegel domain is known to be isomorphic to a nonsymmetric homogeneous bounded domain in $\mathbb{C}^4$ (A proof of this fact will be given in § 9).

§ 4. Generalised Siegel domains

From this section to § 7 we shall follow mostly the work of Kanp - Matsushima - Ochiai [1].

Definition. A domain D in $\mathbb{C}^{n+m}$ is said to be a *generalised Siegel domain* is the following conditions are satisfied.

1) D is isomorphic to a bounded domain in $\mathbb{C}^{n+m}$ and $D \cap (\mathbb{C}^n \times \{0\}) \neq \emptyset$.

2) D is stable under the following affine transformations of $\mathbb{C}^{n+m}$

a) $(z,w) \longrightarrow (z+a,w)$ for any $a \in \mathbb{R}^n$

b) $(z,w) \longrightarrow (z,e^{it}w)$ for any $t \in \mathbb{R}$

c) $(z,w) \longrightarrow (e^t z, e^{ct} w)$ for any $t \in \mathbb{R}$

where c is a real number, called exponent of D, depending only on D.

Examples. In the following, let Ω be an open convex cone in $\mathbb{R}^n$ not containing any straight line.

1) The tube domain $D(\Omega) = \mathbb{R}^n + i\Omega$ is a generalised Siegel domain. Here the exponent c can be taken to be any real number, but we usually take c to be $\frac{1}{2}$.

2) A Siegel domain of the second kind $D(\Omega,F)$ is a generalised Siegel domain with exponent $c = \frac{1}{2}$ by Propositions 1.2 and 2.2.

3) A circular domain D in $\mathbb{C}^m$ is a generalised Siegel domain with $n = 0$ and $c = 0$. Here a domain D is said to be circular if $e^{it}w$ belongs to D for any t in $\mathbb{R}$ whenever $w \in D$.

4) Let $F : \mathbb{C}^m \longrightarrow \bar{\Omega}$ be a map such that

$$F(\lambda w) = |\lambda|^{1/c} . F(w)$$

for some fixed real number $c \neq 0$. Then the set

$$D = \left\{ (z,w) \in \mathbb{C}^{n+m} \mid \operatorname{Im} z - F(w) \in \Omega \right\}$$

is a generalised Siegel domain with exponent c.

Remark. Let Γ be a subgroup of $G(D)$, we say that Γ sweeps D if there exists a compact subset K of D such that $\Gamma K = D$. Then V e y [3] shows the following: Let D be a generalised Siegel domain of exponent c. Suppose that there exists a subgroup Γ of $G(D)$ sweeping D. Then

i) if $c \neq 0$ D is a Siegel domain and

ii) if $c = 0$ D is the direct product $D_1 \times D_2$, where D_1 is a tube domain in $\mathbb{C}^n$ and D_2 is a homogeneous circular domain containing the origin in $\mathbb{C}^m$.

We now list some useful results (mainly due to H. Cartan) about bounded domains in $\mathbb{C}^N$. (See Bochner-Martin, Several complex variables, Princeton University Press).

Let D be a bounded domain in $\mathbb{C}^N$ and $G(D)$ the group of holomorphic automorphisms of D. Then

A_7 $G(D)$ has a structure of real Lie group and the isotropy subgroup of $G(D)$ at a point of D is compact. Moreover $G(D)$ acts on D properly (as a transformation group). We recall that a group G acts properly on D if the map ρ of $G \times D$ into $D \times D$ defined by $\rho(g,x) = (g(x),x)$ is a proper map (i.e. inverse image of any compact set is compact). In particular if $x_0 \in D$ and K is a compact subset of D then $\{ g \in G(D) \mid g(x_0) \in K \}$ is a compact subset of $G(D)$.

B_7 Let x_0 be a point of D. If $g_0 \in G(D)$ is such that $g_0(x_0) = x_0$ and that the linear transformation in the tangent space at x_0 induced by the action of g_0 is identity, then g_0 is the identity element of $G(D)$.

C_7 A vector field X on D is said to be complete if it generates a global one-parameter group of diffeomorphisms $\{ \phi_t \}_{t \in \mathbb{R}}$ of D. On the other hand, if X is a vector field on D and if a local one-parameter group of transformations generated by X consists of holomorphic transformations, X is said to be holomorphic ("par abus de langage", because X is a real vector field and is usually said to be conformal).

Let $\mathfrak{g}(D)$ be the Lie algebra of the Lie group $G(D)$; here we consider $\mathfrak{g}(D)$ as the Lie algebra of all right-invariant vector fields on $G(D)$ (which is, of course, isomorphic to the Lie algebra of all left-invariant vector fields on $G(D)$). Then an element $X \in \mathfrak{g}(D)$ determines a unique one-parameter subgroup in $G(D)$, namely a global one-parameter group of holomorphic automorphisms of D, and hence X determines a unique (real) vector field, say X', on D. It is known and is easily seen that we get thus an isomorphism of the Lie algebra $\mathfrak{g}(D)$ with the Lie algebra of all complete holomorphic vector fields on D. We shall identify $\mathfrak{g}(D)$ with this latter Lie algebra.

We now introduce a formal notation for designing holomorphic vector fields on D. Let X be a holomorphic vector field on D. Denote by $H(D)$ the algebra of all holomorphic functions on D. Then X defines a derivation of the algebra $H(D)$. Moreover since $X(z^A) = X(x^A) + i\,X(y^A)$ $(A = 1,\ldots,N)$ $\{x^A, y^A\}_{A=1,\ldots,N}$ form a real coordinate system on D, X is completely determined by the holomorphic functions $X(z^A)$. Let $X(z^A) = p^A$. Then the derivation of $H(D)$ defined by vector field $\sum_{A=1}^{N} p^A \frac{\partial}{\partial z^A}$ in $H(D)$ agrees with that by X. Thus $\sum_{A=1}^{N} p^A \frac{\partial}{\partial z^A}$ may be used as a formal notation for X and we write $X = \sum_{A=1}^{N} p^A \frac{\partial}{\partial z^A}$ conventionally (although $X = \sum_{A=1}^{N} p^A \frac{\partial}{\partial z^A} + \sum_{A=1}^{N} \bar{p}^A \frac{\partial}{\partial \bar{z}^A}$ in the standard notation).

With this notation, it is easy to verify that if $X = \sum_{1}^{N} p^A \frac{\partial}{\partial z^A}$ and $Y = \sum_{1}^{N} q^A \frac{\partial}{\partial z^A}$ Then,

$$[X,Y] = \sum_{B=1}^{N} \sum_{A=1}^{N} \left(p^A \frac{\partial q^B}{\partial z^A} - q^A \frac{\partial p^B}{\partial z^A}\right) \frac{\partial}{\partial z^B}$$

D_7 Let $X \in \mathfrak{g}(D)$ and suppose that $X_{x_o} = 0$ at a point x_o in D. If $X = \sum_{A=1}^{N} p^A \frac{\partial}{\partial z^A}$ and if the linear part of the Taylor expansion of p^A around x_o vanishes for every A, then $X = 0$. This follows from B_7.

E 7 Let $X = \sum_{A=1}^{N} p^A \frac{\partial}{\partial z^A}$ be a holomorphic vector field on D and define the vector field iX by

$$iX = \sum_{A=1}^{N} ip^A \frac{\partial}{\partial z^A}$$

If X and iX are both in $\mathfrak{g}(D)$, then X = 0.

Let D be a generalised Siegel domain in $\mathbb{C}^{n+m}$. Then

1) $(z,w) \longrightarrow (z + ta, w)$ $(t \in \mathbb{R})$, where $a \in \mathbb{R}^n$

2) $(z,w) \longrightarrow (z, e^{it}w)$ $(t \in \mathbb{R})$

3) $(z,w) \longrightarrow (e^t z\ e^{ct}w)$ $(t \in \mathbb{R})$

are all one parameter subgroups in the group G(D) and hence determine elements of $\mathfrak{g}(D)$. Taking $a = (0,\dots,\overset{k}{1},\dots,0)$ in $\mathbb{R}^n$, the one parameter group 1) defines the element $\frac{\partial}{\partial z^k}$.

Let $\partial_k = \frac{\partial}{\partial z^k}$ $(k = 1,\dots,n)$ and $\partial_\alpha = \frac{\partial}{\partial z^\alpha}$ $(\alpha = n+1,\dots,n+m)$. Let ∂' and ∂ be the elements of $\mathfrak{g}(D)$ determined by the one-parameter groups 2) and 3) respectively. Then we see

$$\partial' = i \sum_{\alpha=n+1}^{n+m} z^\alpha \partial_\alpha \tag{4.1}$$

and

$$\partial = \sum_{k=1}^{n} z^k \partial_k + c \sum_{\alpha=n+1}^{n+m} z^\alpha \partial_\alpha \tag{4.2}$$

It is easily verified that the following relations hold.

Suppose $X = \sum_{k=1}^{n} p^k \partial_k + \sum_{\alpha=n+1}^{n+m} p^\alpha \partial_\alpha$ belongs to $\mathfrak{g}(D)$. Then

$$[\partial_k, X] = \sum_{A=1}^{n+m} \frac{\partial p^A}{\partial z^k} \partial_A \tag{4.3}$$

$$(4.4)\quad [\partial',X] = \sum_{A=1}^{n+m} \partial'(p^A)\,\partial_A - i \sum_{\alpha=n+1}^{n+m} p^\alpha \partial_\alpha$$

$$(4.5)\quad [\partial,X] = \sum_{A=1}^{n+m} \partial(p^A)\,\partial_A - \sum_{k=1}^{n} p^k \partial_k - c \sum_{\alpha=n+1}^{n+m} p^\alpha \partial_\alpha$$

In particular

$$(4.6)\quad \begin{array}{ll} [\partial', \partial_k] = 0 & [\partial', \partial_\alpha] = - i\,\partial_\alpha \\ [\partial, \partial_k] = - \partial_k & [\partial, \partial_\alpha] = - c\,\partial_\alpha \end{array}$$

§ 5. Vector fields belonging to $\mathfrak{g}(D)$

In what follows the indices $j,k,\ell,\ldots$ will run over the integers $1,2,\ldots n$; $\alpha,\beta,\gamma,\ldots$ over $n+1,\ldots n+m$, and $A, B,\ldots$ will stand for integers $1,2,\ldots n+m$. We write ∂_A for $\dfrac{\partial}{\partial z^A}$.

Defintion. Let X be a holomorphic vector field on D and put $X = \sum_{A=1}^{n+m} p^A \partial_A$ where $p^A = X(z^A)$ are holomorphic functions on D. X is said to be a polynomial vector field if the functions p^A are all polynomials of $z^1 \ldots\ldots z^{n+m}$.

Theorem 5.1. Any vector field $X = \sum_{A=1}^{n+m} p^A \partial_A$ belonging to $\mathfrak{g}(D)$ is a polynomial vector field. Moreover, for each fixed $(z^1,\ldots,z^n)$, p^k (respectively p^α) are polynomials of $z^{n+1},\ldots,z^{n+m}$; of total degree ≤ 1 (respectively ≤ 2).

We first prove the following

Lemma 5.1. Let $X = \sum_{A=1}^{n+m} p^A \partial_A \in \mathfrak{g}(D)$. Then the second assertion of Theorem 5.1 holds.

<u>Proof</u>. We have to prove that for fixed $z_o = (z_o^1,\ldots,z_o^n)$, p^k (respectively p^α) are polynomials of $z^{n+1},\ldots z^{n+m}$ of total degree ≤ 1 (respectively ≤ 2). Let $w_o = (z_o^{n+1},\ldots,z_o^{n+m})$. If p^A is defined in a neighbourhood of (z_o,w_o), then in some neighbourhood of w_o it is developed in a Taylor series in $z^{n+1}-z_o^{n+1},\ldots,z^{n+m}-z_o^{n+m}$. It follows that p^A is a polynomial in $z^{n+1},\ldots,z^{n+m}$ of total degree $\leq r$ in a neighbourhood of w_o if and only if $\dfrac{\partial^{r+1} p^A}{\partial z^{\alpha_1}\ldots\partial z^{\alpha_{r+1}}} = 0$ at w_o for any $\alpha_1,\alpha_2,\ldots,\alpha_{r+1}$.

Now, since D is a generalised Siegel domain there exists $z \in \mathbb{C}^n$ such that $(z,o)\in D$. Let $p^A = \sum_{\nu=1}^{\infty} p_\nu^A$ be the Taylor expansion of p^A as function of $z^{n+1},\ldots,z^{n+m}$ around the origin o with the fixed $z = (z^1,\ldots,z^n)$, where p_ν^A is a homogeneous polynomial of $z^{n+1},\ldots,z^{n+m}$ of degree ν. It is easily checked that

$$\sum_{\alpha=n+1}^{n+m} z^\alpha\, \partial_\alpha(p_\nu^A) = \nu . p_\nu^A$$

Now by (4.1) and (4.4),

$$\begin{aligned}(\operatorname{ad}\partial')\,X &= \sum_{A=1}^{n+m} \partial'(p^A)\,\partial_A - i\sum_\alpha p^\alpha\,\partial_\alpha \\ &= \sum_{A=1}^{n+m}\left\{ i\sum_\alpha z^\alpha\,\partial_\alpha\Big(\sum_{\nu=1}^{\infty} p_\nu^A\Big)\right\}\partial_A - i\sum_\alpha p^\alpha\partial_\alpha \\ &= \sum_A\left\{\sum_\nu\Big(\sum_\alpha i\, z^\alpha .\ \partial_\alpha(p_\nu^A)\Big)\right\}\partial_A - i\sum_\alpha p^\alpha\,\partial_\alpha \\ &= \sum_A\Big(\sum_\nu i\,\nu\, p_\nu^A\Big)\,\partial_A - i\sum_\alpha p^\alpha\,\partial_\alpha\end{aligned}$$

$$= \sum_k \Big(\sum_\nu i\nu p_\nu^k \Big) \partial_k + \sum_\alpha \Big(\sum_\nu i(\nu - 1)\, p_\nu^\alpha \Big) \partial_\alpha$$

Taking $(\mathrm{ad}\ \partial')\ X$ for X, we get

$$(\mathrm{ad}\ \partial')^2\ X = \sum_k \Big(\sum_\nu (i\nu)^2 p_\nu^k \Big) \partial_k + \sum_\alpha \Big(\sum_\nu (i(\nu - 1))^2\ p_\nu^\alpha \Big) \partial_\alpha$$

Hence for any real polynomial ϕ with $\phi(0) = 0$ we get

$$(\phi(\mathrm{ad}\ \partial'))\ X = \sum_k \Big(\sum_\nu \phi(i\nu)\ p_\nu^k \Big) \partial_k + \sum_\alpha \Big(\sum_\nu \phi(i(\nu - 1))\ p_\nu^\alpha \Big)$$

Taking in particular $\phi(x) = (x^2 + 1)\, x$

$$\phi(\mathrm{ad}\ \partial')\ X =$$

$$\sum_k \Big(\sum_\nu i\nu(1 - \nu^2)\ p_\nu^A \Big) \partial_k + \sum_\alpha \Big(\sum_\nu i(\nu - 1)(-\nu^2 + 2\nu)\ p_\nu^\alpha \Big) \partial_\alpha .$$

Therefore the vector field $\phi(\mathrm{ad}\ \partial')\ X$ vanishes at (z,o) and linear parts of the components of $\phi(\mathrm{ad}\ \partial')\ X$ at (z,o) also vanish. Then by D) in § 4 it follows that $\phi(\mathrm{ad}\ \partial')\ X = 0$ i.e. $\phi(i\nu)\ p_\nu^k = 0$ and $\phi(i(\nu - 1))\ p_\nu^\alpha = 0$ for all ν. Since $\phi(i\mu) \neq 0$ for $\mu \geq 2$ we get

$$p_\nu^k = 0 \quad \text{for } \nu \geq 2$$

and

$$p_\nu^\alpha = 0 \quad \text{for } \nu \geq 3$$

Now D being open in $\mathbb{C}^{n+m}$, this is also true at (z',o) where z' varies in a neighbourhood of z. Therefore in a neighbourhood of (z,o) in D we get

$$\frac{\partial^2 p^k}{\partial z^{\alpha_1}\, \partial z^{\alpha_2}} = 0$$

$$\frac{\partial^3\, p^\alpha}{\partial z^{\alpha_1} \partial z^{\alpha_2} \partial z^{\alpha_3}} = 0$$

for all $\alpha_1, \alpha_2, \alpha_3$. Since left hand side of these relations are holomorphic functions on D, it follows that they hold everywhere in D. This shows that p^k (respectively p^α) for fixed $z = (z^1,\dots,z^n)$, is a polynomial of $z^{n+1},\dots,z^{m+n}$ of degree ≤ 1 (respectively ≤ 2)

(Q.E.D.)

Proof of theorem 5.1.*) Let $X = \sum_{A=1}^{n+m} p^A \partial_A$ be an element of $\mathfrak{g}(D)$. We have to show that X is a polynomial vector field. By Lemma 5.1 p^A's are polynomials in $z^{n+1},\dots,z^{n+m}$, for fixed $z^1,\dots,z^n$. Therefore it remains to show that for fixed $z^{n+1},\dots,z^{n+m}$, p^A's are polynomials in $z^1, z^2,\dots,z^n$.

For this we consider the vector fields $\partial, \partial_1,\dots,\partial_n$ defined in § 4. Let V denote the $\mathbb{R}$-subspace of $\mathfrak{g}(D)$ generated by $\partial, \partial_1,\dots,\partial_n$. Since $[\partial_k, \partial_\ell] = 0$ and $[\partial, \partial_k] = -\partial_k$ by (4.6), V is a Lie subalgebra of $\mathfrak{g}(D)$. Also V is a solvable Lie-algebra as $\mathfrak{a} = [V,V]$ is an abelian ideal of V. Let 'ad' be the adjoint representation of $\mathfrak{g}(D)$ and consider its restriction to V. Since V is solvable, there exists a basis of the complexification of $\mathfrak{g}(D)$ such that if $x \in V$ then ad x is represented by a matrix of the form

$$\begin{pmatrix} * & & * \\ & * & \\ & & \ddots \\ 0 & & * \end{pmatrix}$$ and therefore if $x \in \mathfrak{a}$ then $\text{ad } x = \begin{pmatrix} 0 & * & * \\ & 0 & \\ & & \ddots \\ 0 & & 0 \end{pmatrix}$

Thus ad x is nilpotent for every $x \in \mathfrak{a}$. Since $\partial_k \in \mathfrak{a}$ there exists an integer r such that $(\text{ad } \partial_k)^r = 0$ for all k. Hence

$$(\text{ad } \partial_k)^r X = \sum_A \frac{\partial^r p^A}{(\partial z^k)^r} \partial_A = 0 \quad \text{for all } k$$

---

*) This simple proof is due to M.S. Raghunathan.

Therefore

$$\frac{\partial^r p^A}{(\partial z^k)^r} = 0 \quad \text{for all} \quad k \quad \text{and} \quad A$$

This shows that for fixed $z^{n+1},\dots,z^{n+m}$; p^A's are polynomials in $z^1,\dots,z^n$

(Q.E.D.)

§ 6. Structure of $\mathfrak{g}(D)$ for D with $c = \frac{1}{2}$

We use the results in Theorem 5.1 to define a gradation on the Lie algebra $\mathfrak{g}(D)$, where D is a generalised Siegel domain with exponent $c = \frac{1}{2}$.

We list below some useful formulae which follow directly from (4.1). Let $Z_{\mu\nu}$ and $W_{\mu\nu}$ denote vector fields in $\mathfrak{g}(D)$ having the following forms

$$Z_{\mu\nu} = \sum_k p^k_{\mu\nu}\, \partial_k$$

$$W_{\mu\nu} = \sum_\alpha p^\alpha_{\mu\nu}\, \partial_\alpha$$

where $p^A_{\mu\nu}$ is a polynomial homogeneous of degree μ in $z^1,\dots,z^n$ and homogeneous of degree ν in $z^{n+1},\dots,z^{n+m}$. Then

(6.1)
$$[\partial, Z_{\mu\nu}] = (\mu - 1 + c\nu)\, Z_{\mu\nu}$$
$$[\partial, W_{\mu\nu}] = (\mu + c.\,(\nu - 1))\, W_{\mu\nu}$$

(6.2)
$$[\partial', Z_{\mu\nu}] = i\nu\, Z_{\mu\nu}$$
$$[\partial', W_{\mu\nu}] = i(\nu - 1)\, W_{\mu\nu}$$

From now on, we assume always that $c = \frac{1}{2}$. The formula (6.1) reduces to

$$(6.3)\qquad \begin{aligned} [\partial, Z_{\mu\nu}] &= (\mu + \tfrac{1}{2}\nu - 1)\, Z_{\mu\nu} \\ [\partial, W_{\mu\nu}] &= (\mu + \tfrac{1}{2}\nu - \tfrac{1}{2})\, W_{\mu\nu} \end{aligned}$$

Let now $X \in \mathcal{G}(D)$. By Theorem 5.1, we have

$$X = \sum_{k=1}^{n} p^k_{\mu\nu}\, \partial_k + \sum_{\alpha=n+1}^{n+m} p^{\alpha}_{\mu\nu}\, \partial_{\alpha}$$

where $p^k_{\mu\nu}$ and $p^{\alpha}_{\mu\nu}$ are polynomials homogeneous of degree μ in $z^1, z^2, \dots z^n$; and homogeneous of degree ν in $z^{n+1}, \dots, z^{n+m}$, further $p^k_{\mu\nu} = 0$ for $\nu \geqslant 2$ and $p^{\alpha}_{\mu\nu} = 0$ for $\nu \geqslant 3$.

For each $\mu = -1, 0, 1, 2, 3, \dots$, let

$$(6.4)\qquad \begin{aligned} X_{\mu} &= \sum_{k=1}^{n} p^k_{\mu+1,0}\, \partial_k + \sum_{\alpha=n+1}^{n+m} p^{\alpha}_{\mu,1}\, \partial_{\alpha} \\ X_{\mu+\frac{1}{2}} &= \sum_{k=1}^{n} p^k_{\mu+1,1}\, \partial_k + \sum_{\alpha=n+1}^{n+m} (p^{\alpha}_{\mu+1,0} + p^{\alpha}_{\mu,2})\, \partial_{\alpha} \end{aligned}$$

Thus, for instance

$$X_{-1} = \sum_{k=1}^{n} p^k_{o,o}\, \partial_k$$

$$X_{-\frac{1}{2}} = \sum_{1}^{n} p^k_{o,1}\, \partial_k + \sum_{n+1}^{n+m} p^{\alpha}_{o,o}\, \partial_{\alpha}$$

$$X_{o} = \sum_{1}^{n} p^k_{1,o}\, \partial_k + \sum_{n+1}^{n+m} p^{\alpha}_{o,1}\, \partial_{\alpha}$$

It is easy to see that $X = \sum_{\lambda \in \Lambda} X_\lambda$ where Λ is the set $\left\{\frac{d}{2} ;\ d \text{ integer } \geqslant -2\right\}$. From (6.3) it follows

$$(\mathrm{ad}\, \partial)X_\lambda = [\partial, X_\lambda] = \lambda X_\lambda \quad \text{for all} \quad \lambda \in \Lambda$$

Thus

$$(\mathrm{ad}\, \partial)X = [\partial, \sum_\lambda X_\lambda] = \sum_\lambda \lambda X_\lambda$$

Hence for any real polynomials ϕ with $\phi(o) = 0$ we get

$$\phi(\mathrm{ad}\, \partial)X = \sum_\lambda \phi(\lambda)X_\lambda$$

Since coefficients of X are polynomials, there exists a $\lambda' \in \Lambda$ such that $X_\lambda = 0$ for all $\lambda \geqslant \lambda'$. For each $\lambda_o \neq 0, \lambda_o < \lambda'$, we can find a real polynomial ϕ with $\phi(o) = 0$ such that $\phi(\lambda_o) = 1$ and $\phi(\lambda) = 0$ for all $\lambda < \lambda', \lambda \neq \lambda_o$. Taking this ϕ, we get

$$\phi(\mathrm{ad}\, \partial)X = X_{\lambda_o}$$

It follows that $X_{\lambda_o} \in \mathfrak{g}(D)$ for each $\lambda_o \neq 0$. Then $X_o = X - \sum_{\lambda \neq 0} X_\lambda$ belongs also to $\mathfrak{g}(D)$. This shows that $X \in \mathfrak{g}(D)$ is decomposed as $X = \sum_\lambda X_\lambda$ with $X_\lambda \in \mathfrak{g}(D)$ for all λ. We have so far proved the first half of the following theorem.

<u>Theorem 6.1</u>. Let D be a generalised Siegel domain with exponent $c = \frac{1}{2}$. For each half-integer $\lambda \geqslant -1$, let $\mathfrak{g}_\lambda$ be the set of all vector fields X_λ in the Lie algebra $\mathfrak{g}(D)$ of the form (6.4). Then

1) $\mathfrak{g}(D) = \sum_{\lambda \in \Lambda} \mathfrak{g}_\lambda$ (direct sum as vector space), where $\Lambda = \left\{\frac{d}{2} ;\ d \text{ integer } \geqslant -2\right\}$

2) $\mathfrak{g}_\lambda$ is the eigenspace of ad ∂ belonging to the eigenvalue λ.

3) If $X \in \mathfrak{g}_\lambda$, $Y \in \mathfrak{g}_\sigma$, then $[X,Y] \in \mathfrak{g}_{\lambda+\sigma}$

Proof. It remains only to prove the assertion 3). But this follows at once from 2) since ad ∂ is a derivation of the Lie algebra of $\mathfrak{g}(D)$.

(Q.E.D.)

Proposition 6.1. Let D be a generalised Siegel domain and let $\mathfrak{g}_{-1}$ be the eigenspace of ad ∂ in $\mathfrak{g}(D)$ corresponding to the eigenvalue -1. Then

$$\mathfrak{g}_{-1} = \left\{ \sum_{k=1}^{n} a^k \partial_k \,\middle|\, a^k \in \mathbb{R} \right\}$$

Proof. Since $\partial_1, \ldots, \partial_n \in \mathfrak{g}_{-1}$ it follows that for any $a^1, \ldots, a^n \in \mathbb{R}$ the vector field $\sum_{k=1}^{n} a^k \partial_k$ belongs to $\mathfrak{g}_{-1}$. Thus

$$\left\{ \sum_{k=1}^{n} a^k \partial_k \,\middle|\, a^k \in \mathbb{R} \right\} \subset \mathfrak{g}_{-1}$$

To prove the converse inclusion consider $X_{-1} \in \mathfrak{g}_{-1}$. By definition $X_{-1} = \sum_{k=1}^{n} c^k \partial_k$, where $c^k \in \mathbb{C}$ for all k and $X_{-1} \in \mathfrak{g}(D)$. Let $Y = \sum_{k=1}^{n} (\operatorname{Im} c^k) \partial_k$. Then $Y \in \mathfrak{g}(D)$ by what we have seen. Also

$$iY = X_{-1} - \sum_{k} (\operatorname{Re} c^k) \partial_k \in \mathfrak{g}(D)$$

Hence by E) in § 4, we get $Y = 0$, showing that $c^k \in \mathbb{R}$.

Thus

$$\mathfrak{g}_{-1} = \left\{ \sum_{k=1}^{n} a^k \partial_k \,\middle|\, a^k \in \mathbb{R} \right\}$$

(Q.E.D.)

Proposition 6.2. Let $D = D(\Omega)$ be a tube domain in $\mathbb{C}^n$. For each integer $\lambda \geq -1$, let $\mathfrak{g}_\lambda = \left\{ \sum_{k=1}^{n} p^k_{\lambda+1} \partial_k \,\middle|\, \sum_{k=1}^{n} p^k_{\lambda+1} \partial_k \in \mathfrak{g}(D) \right\}$ where $p^k_{\lambda+1}$ denotes a homogeneous polynomial of degree $\lambda+1$. Then

1) $\mathfrak{g}(D) = \sum_\lambda \mathfrak{g}_\lambda$

2) $\mathfrak{g}_\lambda$'s are eigenspaces of $\operatorname{ad} \partial$ with eigenvalues λ and $p^k_{\lambda+1}$ are all polynomials with real coefficients.

3) $[\mathfrak{g}_\lambda, \mathfrak{g}_\mu] \subset \mathfrak{g}_{\lambda+\mu}$

Proof. We have only to prove that polynomials in $\mathfrak{g}_\lambda$ are all with real coefficients (other things follow from Theorem 6.1).

$$\text{Let } X = \sum_{k=1}^{n} p^k_{\lambda+1} \partial_k \in \mathfrak{g}_\lambda$$

where

$$p^k_{\lambda+1} = \sum_{j_0 \leq j_1 \leq \cdots \leq j_\lambda} a^k_{j_0 j_1 \cdots j_\lambda} z^{j_0} \cdots z^{j_\lambda}$$

It is easy to see that

$$(\operatorname{ad} \partial_{j_0})(\operatorname{ad} \partial_{j_1}) \cdots (\operatorname{ad} \partial_{j_\lambda}) X = \sum_{k=1}^{n} a^k_{j_0, \cdots j_\lambda} \partial_k$$

Since $(\operatorname{ad} \partial_{j_0}) \cdots (\operatorname{ad} \partial_{j_\lambda}) X \in \mathfrak{g}_1$, it follows from Proposition 6.1 that $a^k_{j_0 \cdots j_\lambda} \in \mathbb{R}$ for all k. Since this is true for any $j_0 \leq j_1 \leq \cdots \leq j_\lambda$, we get that $p^k_{\lambda+1}$ are all real polynomials.

(Q.E.D.)

We now return back to the notation in Theorem 6.1 and shall prove that $\mathfrak{g}_\lambda = (0)$ for all $\lambda \geq \frac{3}{2}$. In other words, we show

$$\mathfrak{g}(D) = \mathfrak{g}_{-1} + \mathfrak{g}_{-\frac{1}{2}} + \mathfrak{g}_{0} + \mathfrak{g}_{\frac{1}{2}} + \mathfrak{g}_{1}$$

Also, we show that the radical $\mathfrak{r}$ of $\mathfrak{g}(D)$ is a graded ideal such that $\mathfrak{r}_\lambda = (0)$ for $\lambda \geq \frac{1}{2}$.

Lemma 6.1. Let D be a generalised Siegel domain with $c = \frac{1}{2}$. Let $\mathfrak{r}$ be the radical of the Lie algebra $\mathfrak{g}(D)$ of the Lie group G(D). Then $\mathfrak{r}$ is a graded ideal of $\mathfrak{g}(D)$ i.e.

$$\mathfrak{r} = \sum_{\lambda \in \Lambda} \mathfrak{r}_\lambda$$

where $\mathfrak{r}_\lambda = \mathfrak{g}_\lambda \cap \mathfrak{r}$. Moreover $\mathfrak{g}_\lambda = \mathfrak{r}_\lambda$ for $\lambda \geq \frac{3}{2}$

Proof. To prove that $\mathfrak{r}$ is a graded ideal, we have to prove that if $X \in \mathfrak{r}$, $X = \sum_{\lambda \in \Lambda} X_\lambda$ with $X_\lambda \in \mathfrak{g}_\lambda$ then $X_\lambda \in \mathfrak{r}$ for all λ. But this follows from the proof of the fact $X_\lambda \in \mathfrak{g}(D)$ made before the statement of Theorem 6.1, if we note that $(\operatorname{ad} \partial)\mathfrak{r} \subset \mathfrak{r}$.

To prove the second part let $X \in \mathfrak{g}_\lambda$, $\lambda \geq \frac{3}{2}$ and for any μ take $Y \in \mathfrak{g}_\mu$ so that $\lambda + \mu \geq \frac{1}{2} > 0$. Then the map ad X ad Y takes any $\mathfrak{g}_\nu$ into $\mathfrak{g}_{(\lambda+\mu)+\nu}$. Therefore ad X ad Y is a nilpotent endomorphism of $\mathfrak{g}(D) = \sum_{\nu \in \Lambda} \mathfrak{g}_\nu$. Let B denote the Killing form of $\mathfrak{g}(D)$. Then it follows

$$B(X,Y) = \operatorname{Tr}(\operatorname{ad} X \operatorname{ad} Y) = 0.$$

Therefore X is orthogonal to any $Y \in \mathfrak{g}_\mu$ with respect to the Killing form B. As μ is arbitrary, X is orthogonal to $\mathfrak{g}(D)$ with respect to B. It follows that X belongs to the radical. (cf. Bourbaki, Groupes et Algebres de Lie I). Therefore $\mathfrak{g}_\lambda = \mathfrak{r}_\lambda$ for $\lambda \geq \frac{3}{2}$.

(Q.E.D.)

Let now $a = (a^1,\dots,a^n) \in \mathbb{R}^n$ and put $A = \sum_{k=1}^{n} a^k \partial_k$.

Clearly $A \in \mathfrak{g}_{-1}$. Define the maps $\phi_A : \mathfrak{g}_1 \longrightarrow \mathfrak{g}_{-1}$ and $\psi_A : \mathfrak{g}_{\frac{1}{2}} \longrightarrow \mathfrak{g}_{-\frac{1}{2}}$

by the formulae :

$$\phi_A(X) = \tfrac{1}{2}(\operatorname{ad} A)^2 X$$

(6.5)

$$\psi_A(X) = (\operatorname{ad} \partial')(\operatorname{ad} A)X$$

Let $X \in \mathfrak{g}_{\frac{1}{2}}$ and write $X = \sum_{k=1}^{n} p^k_{1,1}\,\partial_k + \sum_{\alpha=n+1}^{n+m} (p^\alpha_{1,o} + p^\alpha_{o,2})\,\partial_\alpha$.

Then

$$\begin{aligned}(\operatorname{ad} A)X &= [A, X] \\ &= \sum_{\ell} a^\ell\,[\partial_\ell, X] \\ &= \sum_{k,\ell} a^\ell\,\partial_\ell(p^k_{1,1})\,\partial_k + \sum_{\ell,\alpha} a^\ell\,\partial_\ell(p^\alpha_{1,o})\,\partial_\alpha\end{aligned}$$

Using the formula (6.2) it is observed that

$$(\operatorname{ad} \partial')(\operatorname{ad} A)X = i\sum_{k,\ell} a^\ell\,\partial_\ell(p^k_{1,1})\,\partial_k - i\sum_{\ell,\alpha} a^\ell\,\partial_\ell(p^\alpha_{1,o})\,\partial_\alpha$$

Now suppose a be such that $(ia,o) \in D$. Then

$$\begin{aligned}(\psi_A(X))_{(ia,o)} &= \Big(-i\sum_{\ell,\alpha} a^\ell\,\partial_\ell(p^\alpha_{1,o})\,\partial_\alpha\Big)_{(ia,o)} \\ &= \Big(-i\sum_{\alpha} p^\alpha_{1,o}(a,o)\,\partial_\alpha\Big)_{(ia,o)}\end{aligned}$$

$$= - \sum_{\alpha} p^{\alpha}_{1,o} \, (ia,o) (\partial_{\alpha})_{(ia,o)}$$

$$= - X_{(ia,o)}$$

Thus

(6.6) $$(X + \psi_A(X))_{(ia,o)} = 0$$

for any $X \in \mathfrak{g}_{\frac{1}{2}}$. Similarly for any $X \in \mathfrak{g}_1$, we get

(6.7) $$(X + \phi_A(X))_{(ia,o)} = 0$$

We note that there exists an $a = (a^1 \ldots . a^n) \in \mathbb{R}^n$ such that $(ia,o) \in D$. In fact by definition of generalised Siegel domains there exists a point $(z,o) \in D$, and then $(z - \mathrm{Re}\, z, o) \in D$ so $a = \mathrm{Im}\, z$ satisfies the condition.

It is known that the isotropy subgroup of $G(D)$ at a point (z,w) of D is a compact subgroup of $G(D)$ and the corresponding Lie subalgebra $\mathfrak{k}(z,w)$ of $\mathfrak{g}(D)$ is given by

$$\mathfrak{k}(z,w) = \left\{ X \in \mathfrak{g}(D) \,\middle|\, X_{(z,w)} = 0 \right\}$$

Compactness of the subgroup implies that if $X \in \mathfrak{k}(z,w)$ then ad X is a semisimple endomorphism of $\mathfrak{g}(D)$

<u>Lemma 6.2</u>. Let $\mathfrak{h}$ be a solvable ideal of $\mathfrak{g}(D)$. Then

$$\mathfrak{h} \cap \mathfrak{g}_{\frac{1}{2}} = \mathfrak{h} \cap \mathfrak{g}_1 = (0)$$

<u>Proof</u>. Let $X \in \mathfrak{h} \cap \mathfrak{g}_{\frac{1}{2}}$. Then $\psi_A(X) \in \mathfrak{h} \cap \mathfrak{g}_{-\frac{1}{2}}$ because $\mathfrak{h}$ is an ideal.

Since $\mathfrak{h}^-$ is a solvable Lie algebra we may apply Lie's theorem to the adjoint representation of $\mathfrak{h}^-$ on $\mathfrak{g}(D)^{\mathbb{C}}$, complexification of $\mathfrak{g}(D)$ and conclude that ad X ($X \in \mathfrak{h}^-$) can be expressed simultaneously by upper triangular complex matrices. Since $X \in \mathfrak{g}_{\frac{1}{2}}$, $\psi_A(X) \in \mathfrak{g}_{-\frac{1}{2}}$ it follows that ad X and ad $\psi_A X$ are nilpotent endomorphisms of $\mathfrak{g}(D)$. Therefore ad X and ad $\psi_A(X)$ can be represented simultaneously by upper triangular complex matrices with diagonal zero. Hence ad $(X + \psi_A(X))$ can be represented by an upper triangular complex, matrix with diagonal zero showing that ad $(X + \psi_A(X))$ is nilpotent. On the other hand, $X + \psi_A(X)$ belongs to the subalgebra $\mathfrak{K}(ia,o)$ and therefore ad $(X + \psi_A(X))$ is semisimple. This implies that

$$\text{ad } (X + \psi_A(X)) = 0$$

Therefore

$$[\partial , X + \psi_A(X)] = 0$$

and so

$$\tfrac{1}{2} X + (-\tfrac{1}{2}) \psi_A(X) = 0$$

Now since $X \in \mathfrak{g}_{\frac{1}{2}}$ and $\psi_A(X) \in \mathfrak{g}_{-\frac{1}{2}}$ this implies that $X = 0$

Therefore

$$\mathfrak{h}^- \cap \mathfrak{g}_{\frac{1}{2}} = (0)$$

Similarly using ϕ_A and (6.7) it is proved that

$$\mathfrak{h}^- \cap \mathfrak{g}_1 = (0)$$

(Q.E.D.)

<u>Lemma 6.3</u>. With the same notation as above, $\mathfrak{g}_\lambda = 0$ for $\lambda \geq \frac{3}{2}$

Proof. We first prove that $\mathcal{G}_{\frac{3}{2}} = \mathcal{G}_2 = (0)$. Let $\mathcal{W}$ be the radical in $\mathcal{G}(D)$. We have already shown that $\mathcal{W} = \sum_\lambda \mathcal{W}_\lambda$ and $\mathcal{G}_\lambda = \mathcal{W}_\lambda$ for $\lambda \geq \frac{3}{2}$. Thus $\mathcal{G}_{\frac{3}{2}} \subset \mathcal{W}$. Let $X \in \mathcal{G}_{\frac{3}{2}}$ be given by

$$X = \sum_k p^k_{2,1}\, \partial_k + \sum_\alpha (p^\alpha_{2,0} + p^\alpha_{1,2})\, \partial_\alpha .$$

Then

$$[\partial_\ell, X] = \sum_k (\partial_\ell p^k_{2,1})\, \partial_k + \sum_\alpha (\partial_\ell p^\alpha_{2,0} + \partial_\ell p^\alpha_{1,2})\, \partial_\alpha$$

Now $[\partial_\ell, X] \in \mathcal{G}_{\frac{1}{2}} \cap \mathcal{W} = (0)$ by Lemma 6.2. Hence

$$\partial_\ell p^k_{2,1} = \partial_\ell p^\alpha_{2,0} + \partial_\ell p^\alpha_{1,2} = 0$$

for every ℓ, k, α. Since $p^k_{2,1}, p^\alpha_{2,0}, p^\alpha_{1,2}$ are polynomials homogeneous of degree > 0 in $z^1, \ldots, z^n$, it follows that $p^k_{2,1} = p^\alpha_{2,0} = p^\alpha_{1,2} = 0$. Thus $X = 0$. This proves $\mathcal{G}_{\frac{3}{2}} = (0)$. Similarly it is proved that $\mathcal{G}_2 = (0)$.

We now prove the result by induction on λ. Let $\lambda_0 > 2$. Assume that $\mathcal{G}_\lambda = (0)$ for $\frac{3}{2} \leq \lambda < \lambda_0$. Let $X \in \mathcal{G}_{\lambda_0}$. Then

$$[\partial_k, X] \in \mathcal{G}_{\lambda_0 - 1} = (0)$$

for every k. It follows by the same argument as above that $X = 0$. Thus $\mathcal{G}_{\lambda_0} = (0)$. Since we have already proved that $\mathcal{G}_{\frac{3}{2}} = \mathcal{G}_2 = (0)$

$$\mathcal{G}_\lambda = (0) \quad \text{for all } \lambda \geq \tfrac{3}{2}$$

(Q.E.D.)

Theorem 6.2. Let $\mathfrak{g}(D)$ and $\mathfrak{g}_\lambda$'s be as in Theorem 6.1 and $\mathfrak{h}$ the radical of $\mathfrak{g}(D)$. Then

1) $\mathfrak{g}(D) = \mathfrak{g}_{-1} + \mathfrak{g}_{-\frac{1}{2}} + \mathfrak{g}_0 + \mathfrak{g}_{\frac{1}{2}} + \mathfrak{g}_1$,

2) $\mathfrak{h} = \mathfrak{h}_{-1} + \mathfrak{h}_{-\frac{1}{2}} + \mathfrak{h}_0$, where $\mathfrak{h}_\lambda = \mathfrak{g}_\lambda \cap \mathfrak{h}$,

3) $\dim \mathfrak{g}_{-1} = n$, $\dim \mathfrak{g}_{-\frac{1}{2}} \leq 2m$, $\dim \mathfrak{g}_1 = n - \dim \mathfrak{h}_{-1}$,

$\dim \mathfrak{g}_{\frac{1}{2}} = \dim \mathfrak{g}_{-\frac{1}{2}} - \dim \mathfrak{h}_{-\frac{1}{2}}$.

Proof. The first two parts are already proved in Lemma 6.2 and Lemma 6.3. Also $\dim \mathfrak{g}_{-1} = n$ follows directly from Proposition 6.1. We now prove that $\dim \mathfrak{g}_{-\frac{1}{2}} \leq 2m$. We know that any vector field X in $\mathfrak{g}_{-\frac{1}{2}}$ is of the form

$$X = \sum_k p^k_{o,1} \partial_k + \sum_\alpha c^\alpha \partial_\alpha$$

where $p^k_{o,1}$ are linear forms in $z^{n+1},\dots,z^{n+m}$, and $c = (c^\alpha) \in \mathbb{C}^m$.

Consider the $\mathbb{R}$-linear map f of $\mathfrak{g}_{-\frac{1}{2}}$ into $\mathbb{C}^m$ defined by

$$f(X) = c = (c^\alpha)$$

Then $f(X) = 0$ implies that $X = \sum_{k=1}^{n} p^k_{o,1} \partial_k$. Then $[\partial', X] = iX$ by (6.2). Thus $iX \in \mathfrak{g}(D)$. It follows by E) in § 4 that $X = 0$, showing that f is injective. Since dimension of $\mathbb{C}^m$ over $\mathbb{R}$ is $2m$, it follows that $\dim \mathfrak{g}_{-\frac{1}{2}} \leq 2m$.

Next let $\mathfrak{g}'$ be the quotient Lie algebra $\mathfrak{g}(D)/\mathfrak{r}$. Then $\mathfrak{g}'$ is a semisimple Lie algebra. It follows from 1) and 2) that

$$\mathfrak{g}' = \mathfrak{g}'_{-1} + \mathfrak{g}'_{-\frac{1}{2}} + \mathfrak{g}'_{0} + \mathfrak{g}'_{\frac{1}{2}} + \mathfrak{g}'_{1}$$

where $\mathfrak{g}'_\lambda = \mathfrak{g}_\lambda / \mathfrak{r}_\lambda$. It is easily seen that

$$[\mathfrak{g}'_\lambda , \mathfrak{g}'_\mu] \subset \mathfrak{g}'_{\lambda+\mu}$$

Let B' be the Killing form of $\mathfrak{g}'$. For $X \in \mathfrak{g}'$ and $Y \in \mathfrak{g}'_\mu$ with $\lambda + \mu \neq 0$, ad X ad Y is nilpotent so that $B'(X,Y) = 0$. Also $\mathfrak{g}'$ being semisimple, B' is nondegenerate. It follows that B' restricted to $\mathfrak{g}'_{-\lambda} \times \mathfrak{g}'_\lambda$ is nondegenerate, since we already observed $B'(X,Y) = 0$ for $X \in \mathfrak{g}'_\lambda$, $Y \in \mathfrak{g}'_\mu$ $(\lambda + \mu \neq 0)$. This implies that for $\lambda > 0$

$$\dim \mathfrak{g}'_{-\lambda} = \dim \mathfrak{g}'_\lambda = \dim \mathfrak{g}_\lambda$$

i.e.

$$\dim \mathfrak{g}_{-\lambda} - \dim \mathfrak{r}_{-\lambda} = \dim \mathfrak{g}_\lambda$$

Putting $\lambda = \frac{1}{2}$, 1 we get the desired results.

(Q.E.D.)

<u>Theorem 6.3</u>. Under the same notation as in Theorem 6.1

1) $\mathfrak{a}(D) = \mathfrak{g}_{-1} + \mathfrak{g}_{-\frac{1}{2}} + \mathfrak{g}_0$

where $\mathfrak{a}(D)$ is the Lie subalgebra of $\mathfrak{g}(D)$ corresponding to the closed subgroup AF(D) of G(D), affine automorphism group of D.

2) Let $D_0 = D \cap (\mathbf{C}^n \times \{0\})$. Then Lie subalgebra $\mathfrak{g}_{-1} + \mathfrak{g}_0 + \mathfrak{g}_1$

corresponds to the subgroup $\left\{ g \in G(D) \,\middle|\, g(D_o) \subset D_o \right\}$

<u>Proof</u>. It is clear that a vector field $X = \sum_{k,\mu,\nu} p^k_{\mu,\nu} \partial_k + \sum_{\alpha,\mu,\nu} p^\alpha_{\mu,\nu}$ in $\mathfrak{g}(D)$ is defined by a oneparameter group of affine transformation if and only if $p^k_{\mu,\nu} = p^\alpha_{\mu,\nu} = 0$ whenever $\mu + \nu \geqslant 2$. We use this result to prove 1).

Since $\partial \in \mathfrak{a}(D)$ ad ∂ maps $\mathfrak{a}(D)$ into itself. If follows that

$$\mathfrak{a}(D) = \mathfrak{a}_{-1} + \mathfrak{a}_{-\frac{1}{2}} + \mathfrak{a}_o + \mathfrak{a}_{\frac{1}{2}} + \mathfrak{a}_1$$

where $\mathfrak{a}_\lambda = \mathfrak{g}_\lambda \cap \mathfrak{a}$. It is trivial to see that if $X \in \mathfrak{g}$ for $\lambda \leq 0$ then the corresponding polynomials $p^k_{\mu,\nu} = p^\alpha_{\mu,\nu} = 0$ whenever $\mu + \nu \geqslant 2$ and for all k and α. Thus $X \in \mathfrak{a}(D)$, showing that $\mathfrak{g}_\lambda = \mathfrak{a}_\lambda$ for $\lambda \leq 0$. It remains to show that $\mathfrak{a}_1 = \mathfrak{a}_{\frac{1}{2}} = (0)$. By the remark made at the beginning of this proof, it is obvious $\mathfrak{a}_1 = (0)$. Now let $X \in \mathfrak{a}_{\frac{1}{2}} = \mathfrak{a}(D) \cap \mathfrak{g}_{\frac{1}{2}}$. Since $X \in \mathfrak{g}_{\frac{1}{2}}$

$$X = \sum_k p^k_{1,1} \partial_k + \sum_\alpha (p^\alpha_{1,o} + p^\alpha_{o,2}) \partial_\alpha$$

Also since $X \in \mathfrak{a}(D)$ $p^k_{1,1} = p^\alpha_{o,2} = 0$. Therefore

$$X = \sum_\alpha p^\alpha_{1,o} \partial_\alpha$$

Then, we get $[\partial', X] = iX$ by (6.2). Therefore

$$iX \in \mathfrak{g}(D).$$

By E) in § 4, it follows that $X = 0$ showing that $\mathfrak{a}_{\frac{1}{2}} = (0)$. This completes the proof of the first part 1).

Let $\mathfrak{J}$ be the Lie subalgebra of $\mathfrak{g}(D)$ corresponding to the closed subgroup $\{g \in G(D) \mid g(D_o) \subset D_o\}$ of $G(D)$. Then $X \in \mathfrak{J}$ if and only if X is tangent to D_o at any point (z,o) of D_o. In other words

$X = \sum_{k,\mu,\nu} p^k_{\mu,\nu} \partial_k + \sum_{\alpha,\mu,\nu} p^\alpha_{\mu,\nu} \partial_\alpha$ is in $\mathfrak{J}$ if and only if

$p^\alpha_{\mu,\nu}(z,0) = 0$ for any $(z,0) \in D_o$: Hence $p^\alpha_{\mu,\nu} \neq 0$ implies that $\nu > 0$.

Now since $\partial \in \mathfrak{J}$, $\mathfrak{J}$ is a graded subalgebra of $\mathfrak{g}(D)$ i.e. $\mathfrak{J} = \mathfrak{J}_{-1} + \mathfrak{J}_{-\frac{1}{2}} + \mathfrak{J}_o + \mathfrak{J}_{\frac{1}{2}} + \mathfrak{J}_1$, where $\mathfrak{J}_\lambda = \mathfrak{g}_\lambda \cap \mathfrak{J}$. It is clear that $\mathfrak{J}_{-1} = \mathfrak{g}_{-1}$, $\mathfrak{J}_o = \mathfrak{g}_o$, $\mathfrak{J}_1 = \mathfrak{g}_1$.

It remains to show that $\mathfrak{J}_{-\frac{1}{2}} = \mathfrak{J}_{\frac{1}{2}} = (0)$. Let $X \in \mathfrak{g}_{\frac{1}{2}} \cap \mathfrak{J} = \mathfrak{J}_{\frac{1}{2}}$

Then X is of the form

$$X = \sum_k p^k_{1,1} \partial_k + \sum_\alpha p^\alpha_{o,2} \partial_\alpha$$

Hence

$$[\partial', X] = iX \in \mathfrak{g}(D)$$

Therefore $X = 0$ showing that $\mathfrak{J}_{\frac{1}{2}} = (0)$. Similarly we see $\mathfrak{J}_{-\frac{1}{2}} = (0)$.

(Q.E.D.)

<u>Remark 1</u>. In this last proof, we have actually proved the following fact.

If $X = \sum_k p^k_{1,1} \partial_k + \sum_\alpha (p^\alpha_{1,o} + p^\alpha_{o,2})\partial_\alpha$ belongs to $\mathfrak{g}_{\frac{1}{2}}$ and if $p^\alpha_{1,o} = 0$, then $X = 0$.

Remark 2. Compare Proposition 2.3 with Theorem 6.3 in the case where D is a Siegel domain. The two assertions appear contradictory to each other, as both E and ∂ are defined by the same one-parameter group $\{c_t\}_{t\in\mathbb{R}}$ where $c_t(z,w) = (e^t z, e^{\frac{1}{2}t} w)$. This comes from the fact that we consider $\mathfrak{a}(D)$ as Lie algebras of left and right invariant vector fields on the affine automorphism group AF(D) in Proposition 2.3 and Theorem 6.3 respectively.

§ 7. Lie algebras $\mathfrak{g}(D)$ for Siegel domains

We have proved that if D is a generalised Siegel domain with exponent $c = \frac{1}{2}$ then the Lie algebra $\mathfrak{g}(D)$ is the direct sum

$$\mathfrak{g}(D) = \mathfrak{g}_{-1} + \mathfrak{g}_{-\frac{1}{2}} + \mathfrak{g}_o + \mathfrak{g}_{\frac{1}{2}} + \mathfrak{g}_1 \quad .$$

where $\mathfrak{g}_\lambda$, $(\lambda = -1, -\frac{1}{2}, 0, \frac{1}{2}, 1)$ is an eigenspace of ad ∂ with eigenvalue λ. We now characterise $\mathfrak{g}_{-\frac{1}{2}}$ and $\mathfrak{g}_o$ for a Siegel domain D. Note that $\mathfrak{g}_{-1}$ is already known by Proposition 6.1.

Theorem 7.1. Let $D = D(\Omega,F)$ be a Siegel domain in $\mathbb{C}^{n+m}$. Then

1) $\mathfrak{g}_{-\frac{1}{2}}$ consists of the vector field on $\mathbb{C}^{n+m}$ of the following form

$$2i\sum_{k=1}^{n} F^k(w,c)\, \partial_k + \sum_{\alpha=n+1}^{n+m} c^\alpha \partial_\alpha$$

where $c = (c^\alpha) \in \mathbb{C}^m$ and $F(w,c) = (F^1(w,c), \ldots\ldots . F^n(w,c))$.

In particular, $\dim \mathfrak{g}_{-\frac{1}{2}} = 2m$

2) $\mathfrak{g}_0$ consists of all vector fields of the form

$$\sum_{k,j} a_j^k z^j \partial_k + \sum_{\alpha,\beta} b_\beta^\alpha z^\beta \partial_\alpha$$

where

$A = (a_j^k)$ and $B = (b_\beta^\alpha)$ are matrices with the following properties.

i) A is real, $\exp tA \in \operatorname{Aut} \Omega$ for all $t \in \mathbb{R}$.

ii) $AF(w,w') = F(Bw,w') + F(w,Bw')$.

Proof. 1) Let $X \in \mathfrak{g}_{-\frac{1}{2}}$. So $X = \sum_{k=1}^{n} p_{o,1}^k \partial_k + \sum_{\alpha=n+1}^{n+m} c^\alpha \partial_\alpha$,

where $p_{o,1}^k$ are linear functions in $z^{n+1}, \ldots, z^{n+m}$.

Now for $c = (c^\alpha) \in \mathbb{C}^m$ consider the one parameter group $\{q_{tc}\}_{t \in \mathbb{R}}$ in $G(D)$ defined by

$$q_{tc}(z,w) = (z + 2iF(w,tc) + iF(tc,tc),\ w + tc)$$

This gives rise to a vector field X_c given by

$$X_c = 2i \sum_{k=1}^{n} F^k(w,c)\,\partial_k + \sum_{\alpha=n+1}^{n+m} c^\alpha \partial_\alpha$$

Obviously $X_c \in \mathfrak{g}(D)$ and therefore $X_c \in \mathfrak{g}_{-\frac{1}{2}}$. Since the real dimension of $\{X_c \mid c \in \mathbb{C}^m\}$ is not less than $2m$ and since $\dim \mathfrak{g}_{-\frac{1}{2}} \leq 2m$ by Theorem 6.2, it follows that $\mathfrak{g}_{-\frac{1}{2}}$ consists of vector fields of the form

$$X_c = 2i \sum_{k=1}^{n} F^k(w,c)\,\partial_k + \sum_{\alpha=n+1}^{n+m} c^\alpha \partial_\alpha$$

2) Let $X \in \mathfrak{g}_0$. Then X is of the form

$$X = \sum_{k,j} a^k_j \, z^j \partial_k + \sum_{\alpha,\beta} b^\alpha_\beta \, z^\beta \partial_\alpha$$

where a^k_j, $b^\alpha_\beta \in \mathbb{C}$. Let $A = (a^k_j)$, $B = (b^\alpha_\beta)$, (matrices of degree n and m resp.).

For every $t \in \mathbb{R}$, $\exp tX \in GL(D)$ and

$$\exp tX\,(z,w) = (\exp tA\; z,\ \exp tB\; w)$$

i.e.
$$\exp tX = (\exp tA,\ \exp tB)$$

Using the characterization of $GL(D)$ given in Proposition 2.2, it follows that $\exp tA$ is real and is in $\mathrm{Aut}\,\Omega$; moreover

$$\exp tA\; F(w,w) = F(\exp tB\; w,\ \exp t\,B\; w).$$

It follows that

$$\exp tA\; F(w,w') = F(\exp tB\; w,\ \exp t\,B\; w')$$

and differentiations by t at $t = 0$ of both sides implies

$$AF(w,w') = F(Bw,w') + F(w,Bw').$$

Also $\exp tA$ is real implies that A is a real matrix. This proves the second part.

(Q.E.D.)

Corollary. The subalgebra $\mathfrak{g}_0$ of $\mathfrak{g}(D)$ corresponds to the subgroup $GL(D)$ of $G(D)$, $GL(D)$ being the group of all linear transformations of $\mathbb{C}^{n+m}$ preserving D.

Proof. This follows easily from the theorem in view of Proposition 2.2.

Theorem 7.2. Let $D = D(\Omega, F)$ be a Siegel domain in $\mathbb{C}^{n+m}$. Let $a \in \Omega$ so that $(ia,0) \in D$. Denote the isotropy subalgebra at $(ia,0)$ of $\mathfrak{g}(D)$ by $\mathfrak{k}(ia,0)$. Let $A = \sum_{k=1}^{n} a^k \partial_k \in \mathfrak{g}_{-1}$ and

$$\mathfrak{m} = \left\{ X + \psi_A(X) \mid X \in \mathfrak{g}_{\frac{1}{2}} \right\}$$

$$\mathfrak{m}' = \left\{ X + \phi_A(X) \mid X \in \mathfrak{g}_1 \right\}$$

where ψ_A and ϕ_A are maps defined in § 6. Then,

$$\mathfrak{k}(ia,0) = (\mathfrak{g}_0 \cap \mathfrak{k}(ia,0)) + \mathfrak{m} + \mathfrak{m}'$$

Also $\dim \mathfrak{m} = \dim \mathfrak{g}_{\frac{1}{2}}$ and $\dim \mathfrak{m}' = \dim \mathfrak{g}_1$

Proof. We have already proved by (6.5) that if $X \in \mathfrak{g}_{\frac{1}{2}}$ then $(X + \psi_A(X))\,(ia,0) = 0$ and if $X \in \mathfrak{g}_1$ then $(X + \phi_A(X)(ia,0) = 0$. This shows that $\mathfrak{m}$ and $\mathfrak{m}'$ are contained in $\mathfrak{k}(ia,0)$.

Let $Y \in \mathfrak{k}(ia,0)$ and put $Y = Y_{-1} + Y_{-\frac{1}{2}} + Y_0 + Y_{\frac{1}{2}} + Y_1$ where $Y_\lambda \in \mathfrak{g}_\lambda$. Let $Y' = Y_{\frac{1}{2}} + \psi_A(Y_{\frac{1}{2}})$ and $Y'' = Y_1 + \phi_A(Y_1)$, so that $Y' \in \mathfrak{m}$, $Y'' \in \mathfrak{m}'$. Put $Z = Y - (Y' + Y'') = Y_{-1} + Y_{-\frac{1}{2}} + Y_0 - \psi_A(Y_{\frac{1}{2}}) - \phi_A(Y_1)$

Then clearly $Z \in \mathfrak{g}_{-1} + \mathfrak{g}_{-\frac{1}{2}} + \mathfrak{g}_0$ and $Z \in \mathfrak{k}(ia,0)$

Therefore let

$$Z = \sum_k b^k \partial_k + 2i \sum_k F^k(w,c)\, \partial_k + \sum_\alpha c^\alpha \partial_\alpha + \sum_{k,j} a_j^k z^j \partial_k + \sum_{\alpha,\beta} b_\beta^\alpha z^\beta \partial_\alpha$$

Since $Z_{(ia,0)} = 0$, it follows that

$$b^k + i\sum_j a^k_j a^j = 0 \quad \text{for every } k$$

and $c^\alpha = 0$ for every α . a^j, b^k and a^k_j for $j,k=1,\dots,n$ are all real numbers therefore $b^k = 0$ and $\sum_j a^k_j a^j = 0$ for every k.

So

$$Z = \sum_{k,j} a^k_j z^j \partial_k + \sum_{\alpha,\beta} b^\alpha_\beta z^\beta \partial_\alpha ,$$

showing that $Z \in \mathfrak{g}_0$. Hence $Z \in \mathfrak{g}_0 \cap \mathfrak{k}(ia,0)$. Now $Y = Z + Y' + Y''$ where $Y' \in \mathfrak{m}$, $Y'' \in \mathfrak{m}'$ shows that $Y \in (\mathfrak{g}_0 \cap \mathfrak{k}(ia,0)) + \mathfrak{m} + \mathfrak{m}'$.

This proves the first part of the theorem. The second part follows from the fact that the map taking $X \in \mathfrak{g}_{\frac{1}{2}}$ (respectively $X \in \mathfrak{g}_1$) to $X + \psi_A(X) \in \mathfrak{m}$ (respectively $X + \phi_A(X) \in \mathfrak{m}'$) is bijective and real linear; indeed $X + \psi_A(X) = 0$ implies $X = 0$ since $X \in \mathfrak{g}_{\frac{1}{2}}$ and $\psi_A(X) \in \mathfrak{g}_{-\frac{1}{2}}$.

(Q.E.D.)

<u>Theorem 7.3</u>. If $D = D(\Omega, F)$ is a Siegel domain in $\mathbb{C}^{n+m}$. Then D is affinely homogeneous if and only if D is homogeneous.

<u>Proof</u>. If D is affinely homogeneous then clearly it is homogeneous. Hence we have only to prove the converse. So let D be homogeneous. For any $a \in \Omega$, the point $(ia,0)$ belongs to D. By Theorem 7.2,

$$\begin{aligned} \dim \mathfrak{k}(ia,0) &= \dim(\mathfrak{k}(ia,0) \cap \mathfrak{g}_0) + \dim \mathfrak{m} + \dim \mathfrak{m}' \\ &= \dim(\mathfrak{k}(ia,0) \cap \mathfrak{g}_0) + \dim \mathfrak{g}_{\frac{1}{2}} + \dim \mathfrak{g}_1 \end{aligned}$$

where $\mathfrak{k}(ia,0)$ is the isotropy subalgebra of $\mathfrak{g}(D)$ at $(ia,0) \in D$. Let $K(ia,0)$ be the isotropy subgroup of $G(D)^o$. Then, since $G(D)^o/K(ia,0)$ is homeomorphic to D, it follows

$$\begin{aligned}\dim D &= \dim G(D) - \dim K(ia,0)\\ &= \dim \mathfrak{g}(D) - \dim \mathfrak{k}(ia,0)\\ &= \dim (\mathfrak{g}_{-1} + \mathfrak{g}_{-\frac{1}{2}} + \mathfrak{g}_o) - \dim (\mathfrak{k}(ia,0) \cap \mathfrak{g}_o)\\ &= \dim \mathfrak{a}(D) - \dim (\mathfrak{k}(ia,0) \cap \mathfrak{g}_o)\end{aligned}$$

Now $\mathfrak{k}(ia,0) \cap \mathfrak{g}_o$ is the Lie subalgebra corresponding to the subgroup $G_a \cap K(ia,0)$, G_a being the connected component of the affine automorphism group. Therefore,

$$\begin{aligned}\dim D &= \dim G_a - \dim G_a \cap K(ia,0)\\ &= \dim (G_a/(G_a \cap K(ia,0)))\end{aligned}$$

This implies that $(ia,0)$ is an interior point of the orbit of G_a containing $(ia,0)$ and then this orbit is an open set in D. Now, as remarked in § 3, any point of D can be transformed to $(ia,0)$ for some $a \in \Omega$ by an element of $P(D) \subset G_a$. Thus any orbit of G_a contains a point $(ia,0)(a \in \Omega)$, and so is open. D being connected, orbit of G_a is D. This proves that D is affinely homogeneous.

Theorem 7.4. (Vey [3]). Let $D = D(\Omega,F)$ be a Siegel domain. Assume that the Lie algebra $\mathfrak{g}(D)$ of the Lie group $G(D)$ is a unimodular Lie algebra (i.e. for any $X \in \mathfrak{g}(D)$ $\mathrm{Tr}\,\mathrm{ad}\,X = 0$). Then

1) $\mathfrak{g}(D)$ is a semisimple Lie algebra

2) D is affinely homogeneous.

Proof. Since $\partial \in \mathfrak{g}(D)$, $\operatorname{Tr} \operatorname{ad} \partial = 0$. But, by Theorem 6.2 we get

$$\begin{aligned}\operatorname{Tr} \operatorname{ad} \partial &= (-1) \dim \mathfrak{g}_{-1} + (-\tfrac{1}{2}) \dim \mathfrak{g}_{-\frac{1}{2}} + \tfrac{1}{2} \dim \mathfrak{g}_{\frac{1}{2}} + \dim \mathfrak{g}_1 \\ &= - \dim \mathfrak{g}_{-1} - \tfrac{1}{2} \dim \mathfrak{g}_{-\frac{1}{2}} + \tfrac{1}{2} (\dim \mathfrak{g}_{-\frac{1}{2}} - \dim \mathfrak{W}_{-\frac{1}{2}}) \\ &\qquad + (\dim \mathfrak{g}_{-1} - \dim \mathfrak{W}_{-1}) \\ &= - \dim \mathfrak{W}_{-1} - \tfrac{1}{2} \dim \mathfrak{W}_{-\frac{1}{2}}\end{aligned}$$

Thus we get

$$\dim \mathfrak{W}_{-1} + \tfrac{1}{2} \dim \mathfrak{W}_{-\frac{1}{2}} = 0$$

Therefore

$$\mathfrak{W}_{-1} = \mathfrak{W}_{-\frac{1}{2}} = 0$$

We shall prove that $\mathfrak{W}_0 = 0$. Let $X \in \mathfrak{W}_0 = \mathfrak{W}$. Then

$$[X, \mathfrak{g}_{-1}] \subset \mathfrak{W}_{-1} = (0) \quad \text{and} \quad [X, \mathfrak{g}_{-\frac{1}{2}}] \subset \mathfrak{W}_{-\frac{1}{2}} = (0)$$

Since $\mathfrak{g}_{-1} + \mathfrak{g}_{-\frac{1}{2}}$ is the Lie subalgebra corresponding $P(D)$ it follows that the one-parameter group $\{\phi_t\}_{t \in \mathbb{R}}$ determined by X commutes with every element of $P(D)$. Also, since $X \in \mathfrak{g}_0$, $[\partial, X] = 0$ and therefore $\{c_t\}_{t \in \mathbb{R}}$ commutes with the one-parameter group $\{\phi_t\}_{t \in \mathbb{R}}$ determined by X. Then, by Proposition 2.4 ϕ_t equals identity and so $X = 0$. Thus

$\mathcal{W}_0 = (0)$. Since $\mathcal{W} = \mathcal{W}_{-1} + \mathcal{W}_{-\frac{1}{2}} + \mathcal{W}_0$ by Theorem 6.2, this proves $\mathcal{W} = (0)$ and so the Lie algebra $\mathfrak{g}(D)$ is semi-simple.

We next show that D is affinely homogeneous. By Proposition 3.1, it is sufficient to show that $\mu(GL(D))$ acts transitively on Ω, where μ is the canonical homomorphism of GL(D) into Aut Ω. Let $a \in \Omega$. Then $(ia,0) \in D$. Let $A = \sum_{k=1}^{n} a^k \partial_k \in \mathfrak{g}_{-1}$. Consider the map of $\mathfrak{g}_1$ into $\mathfrak{g}_0$ which takes $X \in \mathfrak{g}_1$ to $[A,X] \in \mathfrak{g}_0$. Put

$$X = \sum_{k} p^k_{2,0} \partial_k + \sum_{\alpha,\ell} z^\ell \partial_\ell (p^\alpha_{1,1}) \partial_\alpha$$

Then

$$[A,X] = \sum_{k,\ell} a^\ell \partial_\ell (p^k_{2,0}) \partial_k + \sum_{\alpha,\ell} a^\ell \partial_\ell (p^\alpha_{1,1}) \partial_\alpha$$

Let

$$p^k_{2,0} = \sum_{h,j=1}^{n} c^k_{hj} z^h z^j \qquad \text{with} \quad c^k_{hj} = c^k_{jh} \quad \text{for all } k.$$

$$\partial_\ell p^k_{2,0} = 2 \sum_j c^k_{\ell j} z^j$$

Now

$$[A,X]_{(ia,0)} = \left\{ \sum_{k,\ell} a^\ell \partial_\ell (p^k_{2,0}) \partial_k \right\}_{(ia,0)}$$

$$= 2 \left\{ \sum_{k,\ell,j} a^\ell c^k_{\ell j} (ia^j) \partial_k \right\}_{(ia,0)}$$

$$= -2i \, X_{(ia,0)}$$

Thus, if $[A,X]_{(ia,0)} = 0$, $X_{(ia,0)} = 0$ i.e. $X \in \mathfrak{k}(ia,0)$. In this case, as noted in § 6, ad X is semisimple. On the other hand, since $X \in \mathfrak{g}_1$ ad X is nilpotent. Thus ad $X = 0$, and $X = 0$ because $\mathfrak{g}(D)$ is semisimple. Thus the map which takes $X \in \mathfrak{g}_1$ to $[A,X]_{(ia,0)}$ is injective. This shows that $\dim \left\{ [A,X]_{(ia,o)} \mid X \in \mathfrak{g}_1 \right\} \geqslant \dim \mathfrak{g}_1$.
Since $[A,X] \in \mathfrak{g}_0$, it follows that

$$\dim \left\{ Y_{(ia,0)} \mid Y \in \mathfrak{g}_0 \right\} \geqslant \dim \left\{ [A,X]_{(ia,0)} \mid X \in \mathfrak{g}_1 \right\}$$

$$\geqslant \dim \mathfrak{g}_1 = n.$$

Now, $\mathfrak{g}_0$ is the Lie algebra of the group $GL(D)$ by Corollary to Theorem 7.1. Since any element of $GL(D)$ preserves $i\Omega \times \{0\}$, any element $Y \in \mathfrak{g}_0$ is then tangent to $i\Omega \times \{0\}$ at $(ia,0)$. In particular, the space $\left\{ Y_{(ia,0)} \mid Y \in \mathfrak{g}_0 \right\}$ is contained in the tangent space of $i\Omega \times \{0\}$ at $(ia,0)$. By the above inequality, we see that the tangent space of $i\Omega \times \{0\}$ coincides with $\left\{ Y_{(ia,0)} \mid Y \in \mathfrak{g}_0 \right\}$. This implies that the orbit of the group $GL(D)$ in $i\Omega \times \{0\}$ containing the point $(ia,0)$ has this point as an interior point, and this means that 'a' is an interior point of the orbit of $\mu(GL(D))$ in Ω, which proves that every orbit of $\mu(GL(D))$ in Ω is open. Since Ω is connected, $\mu(GL(D))$ acts then transitively on Ω.

(Q.E.D.)

Remark. We know that the conclusion of Theorem 7.4 implies that D is isomorphic to a symmetric bounded domain in $\mathbb{C}^{n+m}$.

As to the assumptions of Theorem 7.4, we have the following

Proposition 7.1. The Lie algebra $\mathfrak{g}(D)$ of a Siegel domain D is unimodular, if one of the following condition is satisfied.

1) There exists a discrete subgroup Γ in $G(D)$ such that $G(D)/\Gamma$ is compact

2) There exists a discrete subgroup Γ in $G(D)$ for which there exists a compact set K in D such that $\Gamma K = D$.

Proof. 1) implies that $\mathfrak{g}(D)$ is unimodular. In fact, let U be a neighbourhood of the identity e in $G(D)$ such that $\Gamma \cap U = (e)$. Let V be a relatively compact neighbourhood of e in $G(D)$ such that $V^{-1}V \subset U$. Then, denoting by f the canonical projection $G(D) \longrightarrow G(D)/\Gamma$ one sees immediately that f restricted to g V is one-to-one for every $g \in G(D)$. Let μ be the right invariant Haar measure on $G(D)$. Being given by a right invariant differential form of maximal degree, μ determines a measure μ' on $G(D)/\Gamma$. On the other hand, we get $\mu(gA) = \rho(g)\mu(A)$ for any measurable set A in $G(D)$ where ρ is the 1-dimensional real representation of $G(D)$. Since V is an open relatively compact set, $\mu(V)$ is possitive finite, and we have

$$\rho(g) = \frac{\mu(gV)}{\mu(V)}$$

Because $f: G(D) \longrightarrow G(D)/\Gamma$ is one-to-one in gV, $\mu(gV) = \mu'(f(gV)) \leq \mu'(G(D)/\Gamma)$, and, $G(D)/\Gamma$ being compact, $\mu'(G(D)/\Gamma)$ is finite. It follows that $\rho(g)$ is a bounded representation, and so $\rho(g) = 1$ identically. So μ is a two-sided invariant Haar measure, which implies in particular that $\mathfrak{g}(D)$ is a unimodular Lie algebra.

The assertion 2) implies 1). In fact, as we remarked in § 4, the group $G(D)$ acts properly on D. Let $x_0 \in D$ and put

$$H = \left\{ g \in G(D) \,\middle|\, g(x_o) \in K \right\}$$

Then, it follows that H is a compact set in $G(D)$. Since $\Gamma K = D$ we get $G(D) = \Gamma H$. Thus the quotient space $G(D)/\Gamma$ is compact.

§ 8. Lie algebra $\mathfrak{g}(D)$ for homogeneous Siegel domains

Let D be a Siegel domain and let

$$\mathfrak{g}(D) = \mathfrak{g}_{-1} + \mathfrak{g}_{-\frac{1}{2}} + \mathfrak{g}_{0} + \mathfrak{g}_{\frac{1}{2}} + \mathfrak{g}_{1}$$

be the decomposition of $\mathfrak{g}(D)$ given in Theorem 6.2. Then Theorems 7.1 and 7.2 determine $\mathfrak{g}_{-1}$, $\mathfrak{g}_{-\frac{1}{2}}$, $\mathfrak{g}_{0}$. The purpose of this section is to determine $\mathfrak{g}_{\frac{1}{2}}$, and $\mathfrak{g}_{1}$ successively from the structure of so determined $\mathfrak{g}_{-1}$, $\mathfrak{g}_{-\frac{1}{2}}$, $\mathfrak{g}_{0}$ and $\mathfrak{g}_{\frac{1}{2}}$ under the assumption that D is homogeneous. The results here give a new and simplified version of Tanaka's work [2].

As already noticed in the introduction, if D is a bounded domain in $\mathbb{C}^N$, then there exists a volume element v invariant under the action of $G(D)$. Moreover, put

$$v = i^{N^2} K.dz^1 \wedge \ldots . \wedge dz^N \wedge d\bar{z}^1 \wedge \ldots \wedge d\bar{z}^N$$

where $z^1, \ldots, z^N$ are complex co-ordinates in $\mathbb{C}^N$ and K is a positive function. Then

$$h = ds^2 = \sum_{h,j} \frac{\partial^2 \log K}{\partial z^h \partial \bar{z}^j} dz^h \otimes d\bar{z}^j$$

is a Kaehler metric invariant under the action of $G(D)$.

<u>Lemma 8.1</u>. Let D be a homogeneous bounded domain in $\mathbb{C}^N$ and X be

a holomorphic vector field on D. Then $X \in \mathfrak{g}(D)$ (or equivalently, X is complete) if and only if $L_X v = 0$, where L_X denotes the Lie derivation with respect to X.

Proof. Let $X \in \mathfrak{g}(D)$. Then there exists a one parameter group exp tX in G(D), defining X. Since v is invariant under the action of G(D); we get

$$(\exp tX)^* v = v \qquad (t \in \mathbb{R})$$

Therefore,

$$L_X v = \lim_{t \to o} \frac{1}{t} \left\{ (\exp tX)^* v - v \right\} = 0$$

Conversely, suppose that X is holomorphic and $L_X v = 0$. Now, denoting $dz^1 \wedge \ldots \wedge dz^N$ (resp. $d\bar{z}^1 \wedge \ldots \wedge d\bar{z}^N$) by dz (resp. $d\bar{z}$)

$$L_X v = L_X (i^{N^2} K \; dz \wedge d\bar{z})$$

$$= i^{N^2} \left\{ L_X(K) \, dz \wedge d\bar{z} + K \, L_X dz \wedge d\bar{z} + K \, dz \wedge L_X d\bar{z} \right\}$$

Let $X = \sum_{A=1}^{N} f^A \frac{\partial}{\partial z^A}$ in our notation (cf. § 4).

$$L_X(dz^A) = d(L_X z^A) = d(X(z^A)) = df^A = \sum_B (\partial_B f^A) \, dz^B$$

Similarly

$$L_X(d\bar{z}^A) = \sum_B (\bar{\partial}_B \bar{f}^A) \, d\bar{z}^B$$

where ∂_B (resp. $\bar{\partial}_B$) denotes the vector field $\frac{\partial}{\partial z^B}$ (resp. $\frac{\partial}{\partial \bar{z}^B}$).

Thus

$$L_X v = i^{N^2} K \left\{ (L_X K/K) dz \wedge d\bar{z} + \sum_{A=1}^{N} \partial_A f^A \, dz \wedge d\bar{z} + \sum_{A=1}^{N} \bar{\partial}_A \bar{f}^A \, dz \wedge d\bar{z} \right\}$$

$$= i^{N^2} K \left\{ X(\log K) + \sum_A \partial_A f^A + \sum_A \bar{\partial}_A \bar{f}^A \right\} dz \wedge d\bar{z} .$$

And $L_X v = 0$ implies that

$$X(\log K) + \sum_A \partial_A f^A + \sum_A \partial_A \bar{f}^A = 0.$$

Using this we show that $L_X h = 0$, where h is the Kaehler metric given by

$$h = \sum_{A,B=1}^{N} \frac{\partial^2 \log K}{\partial z^A \partial \bar{z}^B} \, dz^A \otimes d\bar{z}^B$$

$$= \sum_{A,B=1}^{N} (\partial_A \bar{\partial}_B \log K) \, dz^A \otimes d\bar{z}^B$$

In fact,

$$L_X h = \sum_{A,B} X(\partial_A \bar{\partial}_B \log K) \, dz^A \otimes d\bar{z}^B$$

$$+ \sum_{A,B} (\partial_A \bar{\partial}_B \log K) L_X(dz^A) \otimes d\bar{z}^B$$

$$+ \sum_{A,B} (\partial_A \bar{\partial}_B \log K) dz^A \otimes L_X(d\bar{z}^B)$$

$$= \sum_{A,B} X(\partial_A \bar{\partial}_B \log K) dz^A \otimes d\bar{z}^B$$

$$+ \sum_{A,B} ([\partial_A, X] \bar{\partial}_B \log K) dz^A \otimes d\bar{z}^B$$

$$+ \sum_{A,B} (\partial_A [\bar{\partial}_B, X] \log K) dz^A \otimes d\bar{z}^B$$

where we used the formulae $[\partial_A, X] = \sum_C (\partial_A f^C) \partial_C$, $[\bar{\partial}_B, X] = \sum_C (\bar{\partial}_B \bar{f}^C) \bar{\partial}_C$.

It follows that

$$L_X h = \sum_{A,B} \Big\{ (X(\partial_A \bar{\partial}_B) + [\partial_A, X] \bar{\partial}_B + \partial_A [\bar{\partial}_B, X]) \log K \Big\} dz^A \otimes d\bar{z}^B$$

$$= \sum_{A,B} \partial_A \bar{\partial}_B (X \log K) dz^A \otimes d\bar{z}^B .$$

Now by the above formula derived from $L_X v = 0$, we get

$$\partial_A \bar{\partial}_B (X \log K) = - \sum_C \partial_A \bar{\partial}_B \partial_C f^C + \partial_A \partial_B \bar{\partial}_C \bar{f}^C$$

$$= - \sum_C \partial_A \partial_C (\bar{\partial}_B f^C) + \bar{\partial}_B \bar{\partial}_C (\partial_A \bar{f}^C)$$

$$= 0$$

since f^C is holomorphic. So we get $L_X h = 0$, by combining the above formulae. We have so far proved that X is a Killing vector field with respect to the matric h. Now, D being a homogeneous Riemann manifold (with the Riemann metric h), D is a complete Riemann manifold. It is known that on a complete Riemann manifold every Killing vector field is complete (see Kobayashi-Nomizu, Foundations of Differential Geometry, Vol.1, Chap. IV, Theorem 2.4). Thus X is complete. Since X is assumed to be holomorphic, the one-parameter group generated by X consists of holomorphic transformations. Thus $X \in \mathfrak{g}(D)$.

(Q.E.D)

Lemma 8.2. Let $D = D(\Omega, F)$ be a homogeneous Siegel domain. The Bergmann volume v is of the following form.

$$v = i^{N^2} (\lambda \circ \Phi) \quad dz \wedge d\bar{z}$$

where $N = n+m$, λ is a positive function on Ω, Φ is the map given by

$$\Phi(z,w) = \operatorname{Im} z - F(w,w)$$

Moreover, if $A \in GL(D)$, then $\lambda(Ax) = |\det A|^{-2}\lambda(X)$, where A acts on by $\mu(A)$, μ being the canonical homomorphism $GL(D) \longrightarrow \operatorname{Aut} \Omega$.

Proof. We see easily that $p_a^*(dz) = dz$ for any $a \in \mathbb{R}^n$ and $q_b^*(dz) = dz$ for any $b \in \mathbb{C}^m$ where p_a and q_b are transformations of D introduced in § 2. It follows that $dz \wedge d\bar{z}$ is invariant under the action of $P(D)$. Then, as v is invariant under the action of $P(D)$ (in fact under that of the group of $G(D)$), and so is the function K.

Since every element of D can be transformed by an element of $P(D)$ to an element of the form $(ia,0)$ with $a = \Phi(z,w) \in \Omega$, K is completely determined by its values on $i\Omega \times \{0\}$. In other words, if we define the function λ on Ω by

$$\lambda(a) = K(ia,0),$$

then $K = \lambda \circ \Phi$. Thus

$$v = i^{N^2}(\lambda \circ \Phi) dz \wedge d\bar{z}$$

Next let $A = (\mu(A), B) \in GL(D)$. Then

$$A^*(dz \wedge d\bar{z}) = |\det A|^2 \, dz \wedge d\bar{z}$$

Since v is invariant under $G(D)$, and $A \in G(D)$, we get

$$A^*(i^{N^2}(\lambda \circ \Phi) \, dz \wedge d\bar{z}) = i^{N^2}(\lambda \circ \Phi) \, dz \wedge d\bar{z}$$

i.e. $\lambda \circ (\Phi \circ A) \, |\det A|^2 \, dz \wedge d\bar{z} = (\lambda \circ \Phi) \, dz \wedge d\bar{z}$

Therefore

$$|\det A|^2 \, \lambda \circ \Phi \circ A = \lambda \circ \Phi$$

or

$$(\lambda \circ \Phi \circ A)(ix, 0) = |\det A|^{-2} \, (\lambda \circ \Phi)(ix, 0)$$

$$\lambda(Ax) = |\det A|^{-2} \, \lambda(x)$$

(Q.E.D.)

Let $D = D(\Omega, F)$ be a homogeneous Siegel domain in $\mathbb{C}^{n+m}$. We shall denote by $\tilde{\mathfrak{g}}_{\frac{1}{2}}$ the set of all polynomial vector fields on D of the form

$$\sum_{k=1}^{n} p^k_{1,1} \, \partial_k + \sum_{\alpha=n+1}^{n+m} (p^\alpha_{1,0} + p^\alpha_{0,2}) \, \partial_\alpha ,$$

where $p^k_{\mu,\nu}$ and $p^\alpha_{\mu,\nu}$ denote arbitrary polynomials of $z^1, \dots, z^n, z^{n+1}, \dots, z^{n+m}$ which is homogeneous of degree μ in $(z^1, \dots, z^n)$ and of degree ν in $(z^{n+1}, \dots, z^{n+m})$. Then

$$\mathfrak{g}_{\frac{1}{2}} \subset \tilde{\mathfrak{g}}_{\frac{1}{2}}$$

by definition. Similarly we define $\tilde{\mathfrak{g}}_1$ to be the set of all polynomial vector fields of the form

$$\sum_k p^k_{2,0}\, \partial_k + \sum_\alpha p^\alpha_{1,1}\, \partial_\alpha$$

Then

$$\mathfrak{g}_1 \subset \tilde{\mathfrak{g}}_1$$

Theorem 8.1. With the notation above let $X \in \tilde{\mathfrak{g}}_{\frac{1}{2}}$. Then $X \in \mathfrak{g}_{\frac{1}{2}}$ if the following condition is satisfied

$$[X,Y] \in \mathfrak{g}_{-\frac{1}{2}} \qquad \text{for every} \qquad Y \in \mathfrak{g}_{-1}$$

(8.1)

$$[X,Y] \in \mathfrak{g}_0 \qquad \text{for every} \qquad Y \in \mathfrak{g}_{-\frac{1}{2}}$$

Proof. We show that if X satisfies (8.1), then $L_X v = 0$ where v is the volume element as introduced before; in this case, X belongs to $\mathfrak{g}(D)$ by Lemma 8.1 and therefore $X \in \mathfrak{g}_{\frac{1}{2}}$. Now,

$$v = i^{N^2}\, K . dz \wedge d\bar{z}$$

Therefore

$$L_X v = i^{N^2}\left\{ (X(K)\; dz \wedge d\bar{z} + K(L_X dz) \wedge d\bar{z} + K\; dz \wedge L_X d\bar{z}) \right\}$$

Since v is invariant under the action of $G(D)$, v is invariant under the action of one parameter group $\exp tY$ defined by an element Y

of $\mathfrak{g}_{-1}$ or $\mathfrak{g}_{-\frac{1}{2}}$;

$$(\exp tY)^* v = v$$

Therefore

$$\lim_{t \to 0} \frac{1}{t} ((\exp tY)^* v - v) = 0$$

i.e.

$$L_Y v = 0$$

Now X satisfies (8.1). Therefore for $Y \in \mathfrak{g}_{-1}$, $[Y,X] \in \mathfrak{g}_{-\frac{1}{2}}$. Hence

$$L_{[Y,X]} v = 0 \quad \text{and} \quad L_Y v = 0$$

Therefore

$$L_Y L_X v = L_Y L_X v - L_X L_Y v = 0$$

showing that $L_X(v)$ is invariant under the one parameter group $\exp tY$. Similarly considering $Y \in \mathfrak{g}_{-\frac{1}{2}}$ we see that $L_X v$ is invariant under the one parameter subgroup $\exp tY$. Thus $L_X v$ is invariant under the action of the group $P(D)$ whose Lie algebra is $\mathfrak{g}_{-1} + \mathfrak{g}_{-\frac{1}{2}}$. Now, we know that, any $(z,w) \in D$ can be transformed to a point $(ia,0)$ $(a \in \Omega)$ by an element of $P(D)$. Thus if $L_X v$ vanishes at all points $(ia,0) \in D$, it follows that $L_X v = 0$. We have

$$X = \sum_k p^k_{1,1} \, \partial_k + \sum_\alpha (p^\alpha_{1,0} + p^\alpha_{0,2}) \, \partial_\alpha$$

Then
$$\begin{aligned} L_X \, dz^k &= d(L_X z^k) \\ &= d\, p^k_{1,1} \end{aligned}$$

$$= \sum_h \frac{\partial p^k_{1,1}}{\partial z^h} dz^h + \sum_\alpha \frac{\partial p^k_{1,1}}{\partial z^\alpha} dz^\alpha$$

$$L_X dz^\alpha = d(L_X z^\alpha)$$

$$= d(p^\alpha_{1,0} + p^\alpha_{0,2})$$

$$= \sum_h \frac{\partial p^\alpha_{1,0}}{\partial z^h} dz^h + \sum_\beta \frac{\partial p^\alpha_{0,2}}{\partial z^\beta} dz^\beta$$

Therefore

$$L_X dz = L_X (dz^1 \wedge \dots \wedge dz^{n+m})$$

$$= \sum_k (dz^1 \wedge \dots \wedge L_X dz^k \wedge \dots \wedge dz^n) \wedge dz^{n+1} \wedge \dots \wedge dz^{n+m}$$

$$+ dz^1 \wedge \dots \wedge dz^n \wedge \sum_\alpha (dz^{n+1} \wedge \dots \wedge L_X dz^\alpha \wedge \dots \wedge dz^{n+m})$$

$$= \left\{ \sum_k \frac{\partial p^k_{1,1}}{\partial z^k} + \sum_\alpha \frac{\partial p^\alpha_{0,2}}{\partial z^\alpha} \right\} dz^1 \wedge \dots \wedge dz^{n+m}$$

Since $\dfrac{\partial p^k_{1,1}}{\partial z^k}$ and $\dfrac{\partial p^\alpha_{0,2}}{\partial z^\alpha}$ are linear in w, they vanish at all points of the form $(ia,0)$. Therefore

$$(L_X dz)_{(ia,0)} = 0$$

Similarly, we get

$$(L_X d\bar{z})_{(ia,0)} = 0$$

It remains to show that $L_X K = X K = 0$ at $(ia,0) \in D$ $(a \in \Omega)$.

Put $A = \sum_k a^k \partial_k \in \mathfrak{g}_{-1}$ and consider the map ψ_A of $\mathfrak{g}_{\frac{1}{2}}$ into $\mathfrak{g}_{-\frac{1}{2}}$ which takes $Y \in \mathfrak{g}_{\frac{1}{2}}$ to $[\partial', [A,Y]] \in \mathfrak{g}_{-\frac{1}{2}}$. We can extend ψ_A on $\tilde{\mathfrak{g}}_{\frac{1}{2}}$ and, since $X \in \tilde{\mathfrak{g}}_{\frac{1}{2}}$ satisfies (8.1), $\psi_A(X) \in \mathfrak{g}_{-\frac{1}{2}}$. Moreover by a direct calculation as in § 6, we see

$$(X + \psi_A(X))_{(ia,0)} = 0$$

Now, the $\mathbb{R}^n$-valued function Φ on D given by $\Phi(z,w) = \operatorname{Im} z - F(w,w)$ is invariant under $P(D)$. It follows that for any $Y \in \mathfrak{g}_{-1}$ or $\mathfrak{g}_{-\frac{1}{2}}$, $Y\Phi = 0$. In particular, since $\psi_A(X) \in \mathfrak{g}_{-\frac{1}{2}}$, we get $\psi_A(X)\Phi = 0$.

Thus

$$(X\Phi)(ia,0) = ((X + \psi_A(X))\Phi)_{(ia,0)} = 0$$

Now by Lemma 8.2, K is of the form $K = \lambda \circ \Phi$. Let $\Phi^k(z,w)$ be the component of $\Phi(z,w) \in \mathbb{R}^n$. Then the differential of the function K is of the form

$$dK = \sum_k \frac{\partial \lambda}{\partial x^k} \, d\Phi^k$$

where $(x^1,\ldots,x^n)$ are coordinates in $\mathbb{R}^n$ (and so in Ω). We get then

$$\begin{aligned} L_X . K &= X.K \\ &= dK(X) \\ &= \sum_k \frac{\partial \lambda}{\partial x^k} \, d\Phi^k(X) \\ &= \sum_k \frac{\partial \lambda}{\partial x^k} \, (X\Phi^k) \end{aligned}$$

It follows that

$$(L_X K)\,(ia,0) = \sum_k \frac{\partial\lambda}{\partial x^k}\,(X\,\Phi^k)\,(ia,0) = 0$$

Thus, as observed above we get $L_X v = 0$, and this completes the proof.

(Q.E.D.)

We have already seen that the Lie subalgebra $\mathfrak{g}_0$ corresponds to the subgroup $GL(D)$ of $G(D)$. Since $GL(D)$ is the group of linear transformations preserving D, any element $X \in \mathfrak{g}_0$ has the following form

$$X = \sum_k \Big(\sum_j a^k_j z^j\Big)\,\partial_k + \sum_\alpha \sum_\beta \big(b^\alpha_\beta z^\beta\big)\,\partial_\alpha .$$

In this case, using the matrices $A = (a^k_j)$, $B = (b^\alpha_\beta)$ we define the trace of X by the formula

$$\operatorname{Tr} X = \operatorname{Tr} A + \operatorname{Tr} B.$$

Theorem 8.2. Let X be a vector field on D belonging to $\tilde{\mathfrak{g}}_1$, i.e. of the form

$$X = \sum_k p^k_{2,0}\,\partial_k + \sum_\alpha p^\alpha_{1,1}\,\partial_\alpha$$

then $X \in \mathfrak{g}_1$ if and only if the following conditions are satisfied

(8.2) $\quad [X,Y] \in \mathfrak{g}_0$ for every $Y \in \mathfrak{g}_{-1}$

$\quad [X,Y] \in \mathfrak{g}_{\frac{1}{2}}$ for every $Y \in \mathfrak{g}_{-\frac{1}{2}}$

(8.3) $\quad \operatorname{Im} \operatorname{Tr} [X, Y] = 0$ for every $Y \in \mathfrak{g}_{-1}$

<u>Proof</u>. The condition (8.2) is clearly necessary in order that $X \in \mathfrak{g}_1$. Now let $X \in \tilde{\mathfrak{g}}_1$ and suppose X satisfy (8.2). By Lemma 8.1, our theorem follows if we prove that (8.3) implies $L_X v = 0$ and conversely. Since X satisfies (8.2), as in the proof of Theorem 8.1, it follows that $L_X v$ is invariant under the group $P(D)$. Therefore $L_X v = 0$ if it vanishes at all points of the form $(ia,0) \in D$ $(a \in \Omega)$. Now

$$v = i^{N^2} K \, dz \wedge d\bar{z}$$

and

$$L_X v = i^{N^2}(L_X K \, dz \wedge d\bar{z} + K \, L_X dz \wedge d\bar{z} + K \, dz \wedge L_X d\bar{z})$$

We can show $(L_X K)(ia,0) = 0$ as in the proof of Theorem 8.1. In fact let $a \in \Omega$ and put $A = \sum_k a^k \partial_k \in \mathfrak{g}_{-1}$. Consider the map ϕ_A of $\mathfrak{g}_1$ defined by

$$\phi_A(Y) = \tfrac{1}{2}[A, [A,Y]]$$

Then since X satisfies (8.2), $\phi_A(X) \in \mathfrak{g}_{-1}$ and we see

$$(X + \phi_A(X))_{(ia,0)} = 0$$

As in the proof of Theorem 8.1, we get $\phi_A(X)\Phi = 0$ and so $(X\Phi)(ia,0) = 0$; it follows then $(L_X K)((ia,0) = 0$.

Put now

$$p^k_{2,0} = \sum_{h,j} a^k_{hj} \, z^h z^j \quad \text{with} \quad a^k_{hj} = a^k_{jh}$$

and

$$p^{\alpha}_{1,1} = \sum_{j,\beta} b^{\alpha}_{j\beta} \, z^j z^\beta .$$

One can see as in the proof of Theorem 8.1

$$L_X(dz) = \left\{ \sum_k \frac{\partial p^k_{2,0}}{\partial z^k} + \sum_\alpha \frac{\partial p^\alpha_{1,1}}{\partial z^\alpha} \right\} dz$$

$$= \left\{ \sum_k \left(2 \sum_h a^k_{hk} z^h\right) + \sum_\alpha \left(\sum_h b^\alpha_{h\alpha} z^h\right) \right\} dz$$

$$= \left\{ \sum_j \left(2 \sum_k a^k_{hk} + \sum_\alpha b^\alpha_{h\alpha}\right) z^h \right\} dz$$

Now, ∂_h being an element of $\mathfrak{g}_{-1}$ for all h, $[\partial_h, X] \in \mathfrak{g}_0$ and we get

$$[\partial_h, X] = \sum_k \frac{\partial p^k_{2,0}}{\partial z^h} \partial_k + \sum_\alpha \frac{\partial p^\alpha_{1,1}}{\partial z^h} \partial_\alpha$$

$$= \sum_k \left(2 \sum_j a^k_{hj} z^j\right) \partial_k + \sum_\alpha \left(\sum_\beta b^\alpha_{h\beta} z^\beta \right) \partial_\alpha$$

Hence, we get

$$\operatorname{Tr} [\partial_h, X] = 2 \sum_k a^k_{hk} + \sum_\alpha b^\alpha_{h\alpha}$$

Therefore the above formula may be written as

$$L_X dz = \left\{ \sum_h (\operatorname{Tr}[\partial_h, X]) z^h \right\} dz$$

Thus we get

$$(L_X dz)_{(ia,0)} = \Big\{\sum_h (\mathrm{Tr}\,[\partial_h, X])\; ia^h\Big\}\, dz$$

$$= i\ \mathrm{Tr}\,[A,X]\, dz$$

Going back to the formula given in the beginning of this proof, we get

$$(L_X v)_{(ia,0)} = i^{N^2}(L_X K)\ (ia,0)\ dz \wedge d\bar{z}$$

$$+\ i^{N^2} K \Big\{(L_X dz) \wedge d\bar{z} + dz \wedge L_X\, d\bar{z}\Big\}_{(ia,0)}$$

$$= i^{N^2} K(ia,0) \Big\{ i\ \mathrm{Tr}\ [A,X]\ dz \wedge d\bar{z} - i\ \overline{\mathrm{Tr}\,[A,X]}\ dz \wedge d\bar{z}\Big\}_{(ia,0)}$$

$$= i^{N^2} K(ia,0)\ 2(\mathrm{Im}\ \mathrm{Tr}\ [A,X])(dz \wedge d\bar{z})_{(ia,0)}$$

Therefore, if (8.3) is satisfied, we get $(L_X v)_{(ia,0)} = 0$ for any $a \in \Omega$ which yields $L_X v = 0$ as remarked in the beginning of this proof. Conversely, if $L_X v = 0$ then, this formula shows that

$$\mathrm{Im}\ \mathrm{Tr}\,[A,X] = 0$$

for any $A = \sum_k a^k \partial_k \in \mathfrak{g}_{-1}$, where $a = (a^1, \ldots, a^n) \in \Omega$. Since Ω is an open set in $\mathbb{R}^n$, it follows

$$\mathrm{Im}\ \mathrm{Tr}\,[Y,X] = 0$$

for any $Y \in \mathfrak{g}_{-1}$, showing that (8.3) holds for $X \in \mathfrak{g}_1$.

(Q.E.D.)

§ 9. Examples

We shall denote by $M(\ell, q, K)$ the space of all $\ell \times q$ matrices over a field K and by $H(\ell, K)$ the space of all symmetric matrices of degree ℓ over K. The set $H^+(\ell, \mathbb{R})$ of all real positive definite symmetric matrices is an open convex cone not containing any straight line in the real vector space $H(\ell, \mathbb{R})$. It is known that the linear automorphism group of this cone $H^+(\ell, \mathbb{R})$ consists of all transformations of the form

$$(9.1) \qquad x \longrightarrow A\,x\,{}^tA\,, \qquad x \in H^+(\ell, \mathbb{R})$$

where A is an element of the general linear group $GL(\ell, \mathbb{R})$ of degree ℓ.

Now let $m \geqslant 0$, $q \geqslant 0$ and put

$$\mathbb{R}^n = H(m+q, \mathbb{R}).$$

with $n = (m+q)(m+q+1)/2$, and

$$\Omega = H^+(m+q, \mathbb{R}).$$

On the other hand, we may identify $\mathbb{C}^m$ with $M(m, 1, \mathbb{C})$ and we define an Ω-hermitian form F on $\mathbb{C}^m$ with values in $\mathbb{C}^n = H(m+q, \mathbb{C})$ as follows.

$$(9.2) \qquad F(w, w') = \begin{pmatrix} \frac{1}{2}(w\,{}^t\overline{w}' + \overline{w}'\,{}^tw) & 0 \\ 0 & 0 \end{pmatrix}$$

for $w, w' \in \mathbb{C}^m$. Thus we have the Siegel domain

$$D = D(\Omega, F)$$

in $\mathbb{C}^{n+m}$. We shall study the Lie algebra $\mathfrak{g}(D)$ for this domain D.

<u>Case A</u> : $m = 0$

In this case D is the tube domain $D(\Omega)$ in the space $\mathbb{C}^n = H(q, \mathbb{C})$, namely

$$D = \left\{ z = x + iy \;\middle|\; z \in H(q, \mathbb{C}),\ \ y > 0 \right\}.$$

This domain, called Siegel's upper half space, is known to be symmetric; moreover the group of automorphisms of D is given by the action of the symplectic group $Sp(q, \mathbb{R})$ defined as follows.

An element $\begin{pmatrix} A & B \\ C & D \end{pmatrix} \in Sp(q, \mathbb{R})$ acts on D by

$$z \longrightarrow (Az + B)(Cz + D)^{-1} .$$

The isotropy group at $z_o = i1_q$ may be identified with the linear group $GL(q, \mathbb{R})$ modulo $\{\pm 1\}$ because its consists of transformations defined by the elements of the form $\begin{pmatrix} A & 0 \\ 0 & {}^tA^{-1} \end{pmatrix}$ of $Sp(q, \mathbb{R})$ with $A \in GL(q, \mathbb{R})$.

Thus the Lie algebra $\mathfrak{g}(D)$ is isomorphic to the Lie algebra

$$\mathfrak{sp}(q, \mathbb{R}) = \left\{ \begin{pmatrix} A & B \\ C & D \end{pmatrix} \;\middle|\; \begin{matrix} B, C \in H(q, \mathbb{R}) \\ D = -{}^tA \end{matrix} \right\}$$

The decomposition

$$\mathfrak{g}(D) = \mathfrak{g}_{-1} + \mathfrak{g}_0 + \mathfrak{g}_1$$

is such that

$$\mathfrak{g}_{-1} = \left\{ \begin{pmatrix} 0 & B \\ 0 & 0 \end{pmatrix} \;\middle|\; B \in H(q, \mathbb{R}) \right\}$$

$$\mathfrak{g}_0 = \left\{ \begin{pmatrix} A & 0 \\ 0 & -{}^tA \end{pmatrix} \;\middle|\; A \in M(q, q, \mathbb{R}) \right\}$$

$$\mathfrak{g}_1 = \left\{ \begin{pmatrix} 0 & 0 \\ C & 0 \end{pmatrix} \;\middle|\; C \in H(q, \mathbb{R}) \right\}$$

In fact, this is easily verified by the definition of these subspaces.

Case B: $m = 1$, $q = 0$.

In this case

$$D = \left\{ (z,w) \in \mathbb{C}^2 \;\middle|\; \operatorname{Im} z - F(w,w) > 0 \right\}$$

with $F(w,w) = |w|^2$. Thus D is nothing but the elementary Siegel domain $\mathcal{E}^2$ (§ 1). Therefore D is isomorphic to the unit open ball B^2 in $\mathbb{C}^2$ and is a symmetric domain. The Lie algebra $\mathfrak{g}(D)$ is known to be a real simple Lie algebra of dimension 8. As for the decomposition

$$\mathfrak{g}(D) = \mathfrak{g}_{-1} + \mathfrak{g}_{-\frac{1}{2}} + \mathfrak{g}_0 + \mathfrak{g}_{\frac{1}{2}} + \mathfrak{g}_1 ,$$

it follows from Theorems 6.2 and 7.1 the followings.

$$\dim \mathfrak{g}_1 = \dim \mathfrak{g}_{-1} = 1$$

$$\dim \mathfrak{g}_{\frac{1}{2}} = \dim \mathfrak{g}_{-\frac{1}{2}} = 2$$

$$\dim \mathfrak{g}_0 = \dim \mathfrak{g}(D) - 6 = 2.$$

It remains to consider the following cases.

Case C I : $m > 0$, $q > 0$.

Case C II: $m > 1$ $q = 0$.

In the following, we shall examine the structure of $\mathfrak{g}(D)$ exclusively for these two cases.

Proposition 9.1. The group of linear automorphisms GL(D) of the domain D consists of the actions of elements (A, B) of the following form

$$A = \begin{pmatrix} a & b \\ o & c \end{pmatrix}, \quad B = \lambda a$$

where $a \in GL(m, \mathbb{R})$, $c \in GL(q, \mathbb{R})$, $b \in M(m, q, \mathbb{R})$, $\lambda \in \mathbb{C}$, $|\lambda| = 1$. This element (A, B) acts on $\mathbb{C}^{n+m}$ by

$$(A, B)(z, w) = (A\, z\, {}^tA, Bw).$$

Proof. By Proposition 2.2, GL(D) consists of elements $(\tilde{A}, B) \in GL(\mathbb{C}^n) \times GL(\mathbb{C}^m)$ such that i) $A \in \mathrm{Aut}\,\Omega$ and ii) $A\, F(w,w) = F(Bw, Bw)$ $(w \in \mathbb{C}^m)$. In our case $\tilde{A} \in \mathrm{Aut}\,\Omega$ is given by

$$x \longrightarrow A\, x\, {}^tA \qquad (x \in \mathbb{R}^n)$$

with $A \in GL(m + q, \mathbb{R})$. Put

$$A = \begin{pmatrix} a & b \\ d & c \end{pmatrix}$$

Then, the condition ii) means that ii)' $A\, F(w, w')\, {}^tA = F(Bw, Bw')$ $(w, w' \in \mathbb{C}^m)$. Since F is defined by (9.2), if (A, B) is of the form in the proposition, (A, B) satisfies i) and ii)' and therefore (A, B) belongs to GL(D). Conversely, suppose (A, B) satisfies ii)' with $A \in GL(m+q, \mathbb{R})$. Then, by exhibiting both sides of ii)' and taking w, w' suitably, we see easily

$d = 0$ in A. Moreover, we get $a\, F(w,w')\, {}^t a = F(Bw, Bw')$ and so

$$F(w,w') = F(a^{-1} Bw,\ a^{-1} Bw')$$

From this relation, taking w, w' suitably, we can conclude easily that $a^{-1} Bw = \lambda w$ for a complex number λ with $|\lambda| = 1$. Therefore $B = \lambda a$ with $|\lambda| = 1$, which proves our proposition.

(Q.E.D)

Let now

$$\mathfrak{g}(D) = \mathfrak{g}_{-1} + \mathfrak{g}_{-\frac{1}{2}} + \mathfrak{g}_0 + \mathfrak{g}_{\frac{1}{2}} + \mathfrak{g}_1$$

be the decomposition of $\mathfrak{g}(D)$ as in Theorem 6.2. By Proposition 6.1 and Theorem 7.1 we know $\mathfrak{g}_{-1}$, $\mathfrak{g}_{-\frac{1}{2}}$. To describe them exactly let k, l, j, h,.... run over $1,\dots\dots,m+q$ and α, β, γ,... over $1,\dots,m$. The elements of $\mathbb{C}^m$ are then written as $w = (w^\alpha)$ and those of $\mathbb{C}^n = H(m+q, \mathbb{C})$ is $z = (z^{kl})$ with $z^{kl} = z^{lk}$. Thus z^{kl} $(k \leq l)$ are co-ordinates of z with double index kl. We make the convention $\partial_{kl} = \partial_{lk} = \partial/\partial z^{kl}$. Then

$$(9.3) \qquad \mathfrak{g}_{-1} = \left\{ \sum_k a^{kl} \partial_{kl} \,\middle|\, a^{kl} \in \mathbb{R} \right\}$$

$$(9.4) \qquad \mathfrak{g}_{-\frac{1}{2}} = \left\{ 2i \sum_{k \leq l} F^{kl}(w,c)\, \partial_{kl} + \sum_\alpha c^\alpha \partial_\alpha \,\middle|\, c \in \mathbb{C}^m \right\}$$

where

$$(9.5) \qquad F^{kl}(w,c) = \begin{cases} \frac{1}{2}\,(w^\alpha \bar{c}^\beta + \bar{c}^\alpha w^\beta), & \text{if } (k,l) = (\alpha,\beta) \\ 0, & \text{if } k \text{ or } l > m. \end{cases}$$

by (9.2). It follows

(9.6) $\quad \partial_\gamma F^{\alpha\beta}(w,c) = \frac{1}{2}(\delta_{\gamma\alpha}\bar{c}^\beta + \bar{c}^\alpha \delta_{\gamma\beta})$

where $\delta_{\gamma\alpha}$ is, of course, Kronecker's symbol.

Now, since $\mathfrak{g}_0$ is the Lie algebra of the group GL(D), it follows from Proposition 9.1.

(9.7) $$\mathfrak{g}_0 = \left\{ (A, B) \,\middle|\, A = \begin{pmatrix} a & b \\ o & c \end{pmatrix}, \quad B = i\, t\, 1_m + a \right\}$$

where a, b, c are real matrices of type $m \times m$, $m \times q$, $q \times q$ respectively and t is any real number.

<u>Proposition 9.2</u>. $\mathfrak{g}_{\frac{1}{2}} = (0)$ for the Cases C I and C II.

<u>Proof</u>. Let $X \in \mathfrak{g}_{\frac{1}{2}}$. Then X is written as

(9.8) $$X = \sum_k p^k_{1,1}\, \partial_k + \sum_\alpha (p^\alpha_{1,0} + p^\alpha_{0,2})\, \partial_\alpha$$

Put

(9.9) $$p^\alpha_{1,0} = \sum_{k \leq \ell} d^\alpha_{k\ell} z^{k\ell} .$$

We have

(9.10) $\quad [Y, X] \in \mathfrak{g}_{-\frac{1}{2}} \qquad$ for $\quad Y \in \mathfrak{g}_{-1}$

(9.11) $\quad [Y, X] \in \mathfrak{g}_0 \qquad$ for $\quad Y \in \mathfrak{g}_{-\frac{1}{2}}$

We shall show that these conditions imply $p^\alpha_{1,0} = 0$ for all α. Then, as remarked at the end of § 6, we get $X = 0$ and $\mathfrak{g}_{\frac{1}{2}} = (0)$ will be proved.

First we get

(9.12) $$[\partial_{k\ell}, X] = \sum_{h \leq j} \partial_{k\ell}\, p^{hj}_{1,1}\, \partial_{hj} + \sum_\alpha d^\alpha_{k\ell}\, \partial_\alpha$$

and this belongs to $\mathfrak{g}_{-\frac{1}{2}}$ by (9.10). By (9.4) we have then for each (k, ℓ, h, j)

$$\partial_{k\ell}\, p^{hj}_{1,1} = 2i\, F^{hj}(w,c)$$

with $c = (d^{\alpha}_{k\ell})$. Because of (9.5) this means

$$(9.13)\qquad \begin{cases} \partial_{k\ell}\, p^{\alpha\beta}_{1,1} = i(w^{\alpha}\, \bar{d}^{\beta}_{k\ell} + \bar{d}^{\alpha}_{k\ell}\, w^{\beta}) \\ \partial_{k\ell}\, p^{hj}_{1,1} = 0 \quad \text{if } h \text{ or } j > m. \end{cases}$$

Therefore

$$(9.14)\qquad \begin{cases} p^{\alpha\beta}_{1,1} = i \sum\limits_{k\le\ell} (w^{\alpha}\, \bar{d}^{\beta}_{k\ell} + \bar{d}^{\alpha}_{k\ell}\, w^{\beta})\, z^{k\ell} \\ p^{hj}_{1,1} = 0 \quad \text{if } h \text{ or } j > m. \end{cases}$$

Now let $Y \in \mathfrak{g}_{-\frac{1}{2}}$. By (9.4) Y can be expressed with a $c \in \mathbb{C}^m$ in the form

$$(9.15)\qquad Y = 2i \sum_{k\le\ell} F^{k\ell}(w,c)\, \partial_{k\ell} + \sum_{\alpha} c^{\alpha}\, \partial_{\alpha}$$

Then $[Y, X] \in \mathfrak{g}_0$ by (9.11). Now

$$(9.16)\quad [Y, X] = 2i \sum_{k\le\ell} \Big[F^{k\ell}(w,c)\, \partial_{k\ell},\ \sum_{h\le j} p^{hj}_{1,1}\, \partial_{hj} + \sum_{\alpha} (p^{\alpha}_{1,0} + p^{\alpha}_{0,2})\, \partial_{\alpha} \Big]$$

$$+ \sum_{k\le\ell} \sum_{\gamma} c^{\gamma}\, \partial_{\gamma} p^{k\ell}_{1,1}\, \partial_{k\ell} + \sum_{\alpha,\gamma} c^{\gamma}\, \partial_{\gamma} p^{\alpha}_{0,2}\, \partial_{\alpha}$$

$$= 2i \sum_{\substack{k\le\ell \\ h\le j}} F^{k\ell}(w,c)\, \partial_{k\ell}\, p^{hj}_{1,1}\, \partial_{hj} + 2i \sum_{k\le\ell,\alpha} F^{k\ell}(w,c)\, \partial_{k\ell}\, p^{\alpha}_{1,0}\, \partial_{\alpha}$$

$$- 2i \sum_{k \leq l, \alpha} (p^{\alpha}_{1,0} + p^{\alpha}_{0,2}) \, (\partial_{\alpha} F^{k l}(w,c)) \, \partial_{k l}$$

$$+ \sum_{k \leq l, \gamma} c^{\gamma} \partial_{\gamma} p^{k l}_{1,1} \partial_{k l} + \sum_{\alpha, \gamma} c^{\gamma} \partial_{\gamma} p^{\alpha}_{0,2} \partial_{\alpha} .$$

By the definition of elements of $\mathfrak{g}_0$, the coefficients of ∂_{hj} of an element of $\mathfrak{g}_0$ is a linear form of (z^{kl}).

It follows from (9.16) that these coefficients of $[Y, X]$ are:

(9.17) $$- 2i \sum_{\alpha} p^{\alpha}_{1,0} \partial_{\alpha} F^{k l}(w,c) + \sum_{\alpha} c^{\alpha} \partial_{\alpha} p^{k l}_{1,1}$$

$$= \begin{cases} 0 \quad \text{if } k \text{ or } l > m \quad \text{(by (9.5), (9.14))} \\[2ex] - 2i \displaystyle\sum_{\alpha} \sum_{h \leq j} d^{\alpha}_{hj} z^{hj} \tfrac{1}{2} (\delta_{\alpha\gamma} \bar{c}^{\epsilon} + \delta_{\alpha\epsilon} \bar{c}^{\gamma}) \\[2ex] \quad + \displaystyle\sum c^{\alpha} \sum_{h \leq j} i \, (\delta_{\alpha\gamma} d^{\epsilon}_{hj} + \delta_{\alpha\epsilon} \bar{d}^{\gamma}_{hj}) \, z^{hj} \\[1ex] \quad \text{if } (k, l) = (\gamma, \epsilon) \quad \text{(by 9.5), (9.9), (9.14)} \end{cases}$$

$$= \begin{cases} 0 \quad \text{if } k \text{ or } l \geq m \\[2ex] 2 \displaystyle\sum_{h \leq j} \mathrm{Im} \, (\bar{c}^{\epsilon} d^{\gamma}_{hj} + \bar{c}^{\gamma} d^{\epsilon}_{hj}) \, z^{hj} \quad \text{if } (k, l) = (\gamma, \epsilon) \end{cases}$$

Consider now Case C I : $m > 0, \quad q > 0$

By (9.7) an element $X_0 \in \mathfrak{g}_0$ is of the form (A, B) with

$A = \begin{pmatrix} a & b \\ o & c \end{pmatrix}$ and X_0 acts on $\mathbb{C}^n = H(m + q, \mathbb{C})$ by

$$\tilde{A}.z = Az + z \, {}^{t}A \qquad (z \in H(m + q, \mathbb{C}))$$

As a vector field on D, X_o is written as

$$(9.18) \qquad X_o = \sum_{\substack{k \leq \ell \\ h \leq j}} A^{k\ell}_{hj} \, z^{hj} \, \partial_{k\ell} + \sum_{\alpha,\beta} B^{\alpha}_{\beta} \, w^{\beta} \, \partial_{\alpha} .$$

Then $(A^{k\ell}_{hj})$ is the matrix representing the linear transformation $\tilde{A}$ with respect to the natural basis of $H(m+q, \mathbb{C})$. Suppose $X_o = [Y, X]$. Then (9.17) shows that

$$A^{k\ell}_{hj} = 0 \quad \text{if} \quad k \text{ or } \ell > m.$$

This means that the transformation $\tilde{A}$ maps the whole space $H(m+q, \mathbb{C})$ into the <u>proper</u> subspace identified with $H(m, \mathbb{C})$, consisting of matrices of the form

$$\begin{pmatrix} a & 0 \\ 0 & 0 \end{pmatrix} \qquad a \in H(m, \mathbb{C}).$$

But, taking $z = \begin{pmatrix} 0 & y \\ {}^t y & w \end{pmatrix}$ with $w \in H(q, \mathbb{C})$, $y \in M(m, q, \mathbb{C})$, we have

$$\tilde{A}(z) = \begin{pmatrix} b\,{}^t y + y\,{}^t b, & ay + bw + y\,{}^t c \\ c\,{}^t y + {}^t y\,{}^t a + w\,{}^t b, & cw + w\,{}^t c \end{pmatrix}$$

It follows from $\tilde{A}(z) \in H(m, \mathbb{C})$ for any $w \in H(q, \mathbb{C})$ and $y \in M(m, q, \mathbb{C})$ that $c = 0$, $a = 0$ and $b = 0$. Thus $A = 0$ and $A^{k\ell}_{hj} = 0$ for all $k\ell$, hj. By (9.17) we have then

$$\operatorname{Im} (\bar{c}^{\epsilon} d^{\gamma}_{hj} + \bar{c}^{\gamma} d^{\epsilon}_{hj}) = 0$$

for all γ, ϵ, h, j. This being true for all $c = (c^{\alpha}) \in \mathbb{C}^m$, putting

$c = (0,..,1,0,..,0)$ or $c = (0,..,i,...,0)$ it follows $\operatorname{Im} d_{hj} = \operatorname{Re} d^{\gamma}_{hj} = 0$ and so $d^{\gamma}_{hj} = 0$ for all γ, h, j. Thus $p^{\alpha}_{1,0} = 0$ which implies $X = 0$ as remarked in the beginning of this proof. Thus we have proved $\mathfrak{g}_{\frac{1}{2}} = (0)$ for the Case C I.

Consider now the Case C II: $m > 1$, $q = 0$.

In this case the formulas (9.8) and (9.9) for $X \in \mathfrak{g}_{\frac{1}{2}}$ are written as

$$X = \sum_{\alpha \leq \beta} p^{\alpha\beta}_{1,1} \partial_{\alpha\beta} + \sum_{\alpha} (p^{\alpha}_{1,0} + p^{\alpha}_{0,2}) \partial_{\alpha}$$

$$p^{\alpha}_{1,0} = \sum_{\gamma \leq \delta} d^{\alpha}_{\gamma\delta} z^{\gamma\delta} .$$

We shall make use of the following fact. If $d^{\alpha}_{\gamma\delta}$ is purely imaginary for a given (α, γ, δ) and for any $X \in \mathfrak{g}_{\frac{1}{2}}$, then $d^{\alpha}_{\gamma\delta} = 0$. In fact, by (6.2) we get

$$[\partial, X] = i \sum_{\alpha \leq \beta} p^{\alpha\beta}_{1,1} \partial_{\alpha\beta} + \sum_{\alpha} (-i\, p^{\alpha}_{1,0} + i\, p^{\alpha}_{0,2}) \partial_{\alpha}$$

and this belongs also to $\mathfrak{g}_{\frac{1}{2}}$. The coefficient of $z^{\gamma\delta}$ for the given (α, γ, δ) of $-i\, p^{\alpha}_{1,0}$ is $-i d^{\alpha}_{\gamma\delta}$ which should be also purely imaginary by assumption. So we get $d^{\alpha}_{\gamma\delta} = 0$.

Now the formula (9.16) is :

$$(9.16)' \quad [Y, X] = 2i \sum_{\substack{\gamma \leq \delta \\ \alpha \leq \beta}} F^{\gamma\delta}(w,c)\, \partial_{\gamma\delta}\, p^{\alpha\beta}_{1,1}\, \partial_{\alpha\beta}$$
$$+ 2i \sum_{\gamma \leq \delta, \alpha} F^{\gamma\delta}(w,c)\, \partial_{\gamma\delta}\, p^{\alpha}_{1,0}\, \partial_{\alpha}$$
$$- 2i \sum_{\alpha \leq \beta, \gamma} (p^{\gamma}_{1,0} + p^{\gamma}_{0,2})\, \partial_{\gamma}\, F^{\alpha\beta}(w,c)\, \partial_{\alpha\beta}$$

$$+ \sum_{\alpha \leq \beta, \gamma} c^{\gamma} \quad \partial_{\gamma} p^{\alpha\beta}_{1,1} \; \partial_{\beta} + \sum_{\alpha,\gamma} c^{\gamma} \partial_{\gamma} p^{\alpha}_{0,2} \partial_{\alpha}$$

Since the coefficient of $\partial_{\alpha\beta}$ must be linear in $z^{\alpha\beta}$ in this formula, it follows:

$$2i \sum_{\gamma \leq \delta} F^{\gamma\delta}(w,c) \, \partial_{\gamma\delta} \; p^{\alpha\beta}_{1,1} - 2i \sum_{\gamma} p^{\gamma}_{0,2} \; \partial_{\gamma} F^{\alpha\beta}(w,c) = 0$$

Applying (9.5) (9.6), (9.13), we get

$$i \sum_{\gamma \leq \delta} (w^{\gamma} \bar{c}^{\delta} + \bar{c}^{\gamma} w^{\delta})(i \, (w^{\alpha} \bar{d}^{\beta}_{\gamma\delta} + \bar{d}^{\alpha}_{\gamma\delta} w^{\beta}))$$

$$- i \sum_{\gamma} p^{\gamma}_{0,2} \; (\delta_{\gamma\alpha} \bar{c}^{\beta} + \bar{c}^{\alpha} \; \delta_{\gamma\beta}) = 0$$

$$p^{\alpha}_{0,2} \; \bar{c}^{\beta} + p^{\beta}_{0,2} \; \bar{c}^{\alpha} = 2i \sum_{\gamma} w^{\gamma} \bar{c}^{\gamma} (w^{\alpha} \bar{d}^{\beta}_{\gamma\gamma} + \bar{d}^{\alpha}_{\gamma\gamma} w^{\beta})$$

$$+ i \sum_{\gamma \neq \delta} w^{\gamma} \bar{c}^{\delta} \; (w^{\alpha} \bar{d}^{\beta}_{\gamma\delta} + \bar{d}^{\alpha}_{\gamma\delta} w^{\beta})$$

where we make the convention that $d^{\alpha}_{\gamma\delta} = d^{\alpha}_{\delta\gamma}$ for $\gamma < \delta$. Put $\alpha = \beta$ and take the case $c = (0,.,\overset{\alpha}{1},..,0)$. Then we get

$$p^{\alpha}_{0,2} = 2i \; \bar{d}^{\alpha}_{\alpha\alpha} \; w^{\alpha} \; w^{\alpha} + i \sum_{\alpha \neq \gamma} \bar{d}^{\alpha}_{\alpha\gamma} w^{\alpha} w^{\gamma} \tag{9.19}$$

Therefore

$$\partial_{\gamma} p^{\alpha}_{0,2} = \begin{cases} 4i \; \bar{d}^{\alpha}_{\alpha\alpha} \; w^{\alpha} + i \sum_{\gamma \neq \alpha} \bar{d}^{\alpha}_{\alpha\gamma} \; w^{\gamma} & \text{if } \gamma = \alpha \\ i \; \bar{d}^{\alpha}_{\alpha\gamma} \; w^{\alpha} & \text{if } \gamma \neq \alpha . \end{cases} \tag{9.20}$$

We shall now consider the coefficient of ∂_{α} in (9.16)', which should be linear in w^{α} because $[Y, X] \in \mathfrak{g}_0$. Thus, it is given by

$$2i \sum_{\mu \le \delta} F^{\mu\delta}(w,c)\, \partial_{\mu\delta}\, p^{\alpha}_{1,0} + \sum_{\alpha,\mu} c^{\mu} (\partial_{\mu} p^{\alpha}_{0,2})$$

$$= i \sum_{\mu \le \delta} (w^{\mu}\bar{c}^{\delta} + \bar{c}^{\mu} w^{\delta})\, d^{\alpha}_{\mu\delta} + i \sum_{\mu \ne \alpha} c^{\mu}\, \bar{d}^{\alpha}_{\alpha\mu}\, w^{\alpha}$$

$$+ 4i\ c^{\alpha}\, \bar{d}^{\alpha}_{\alpha\alpha}\ w^{\alpha} + i \sum_{\mu \ne \alpha} c^{\mu}\, \bar{d}^{\alpha}_{\alpha\mu}\, w^{\mu} \qquad \text{(by (9.20))}$$

$$= 2i \sum \bar{c}^{\mu}\, d^{\alpha}_{\mu\mu}\, w^{\mu} + i \sum_{\mu \ne \delta} \bar{c}^{\mu}\, d^{\alpha}_{\mu\delta}\, w^{\delta}$$

$$+ i \sum_{\mu \ne \alpha} c^{\mu}\, \bar{d}^{\alpha}_{\alpha\mu}\, w^{\alpha} + i \sum_{\mu \ne \alpha} c^{\alpha}\, \bar{d}^{\alpha}_{\alpha\mu} w^{\mu}$$

$$+ 4i\ c^{\alpha}\ \bar{d}^{\alpha}_{\alpha\alpha}\ w^{\alpha}$$

Let b^{α}_{μ} be the coefficient of w^{μ} in this formula. Then, if $\mu \ne \alpha$

$$b^{\alpha}_{\mu} = 2i\ \bar{c}^{\mu}\, d^{\alpha}_{\mu\mu} + i \sum_{\beta \ne \mu} \bar{c}^{\beta}\, d^{\alpha}_{\beta\mu} + i\ c^{\alpha}\ \bar{d}^{\alpha}_{\alpha\mu}$$

and

$$b^{\alpha}_{\alpha} = 2i\ \bar{c}^{\alpha}\, d^{\alpha}_{\mu\mu} + i \sum_{\mu \ne \alpha} (c^{\mu}\bar{d}^{\alpha}_{\alpha\mu} + \bar{c}^{\mu} d^{\alpha}_{\alpha\mu}) + 4i\ c^{\alpha}\ \bar{d}^{\alpha}_{\alpha\alpha}\ .$$

Now if $[Y, X] \in \mathfrak{g}_0$ is represented by (A, B) and if $B = (B^{\alpha}_{\beta})$, then

$$[Y, X] = \sum_{\substack{\alpha \le \beta \\ \mu \le \delta}} A^{\alpha\beta}_{\mu\delta}\ z^{\mu\delta}\ \partial_{\alpha\beta} + \sum_{\alpha} B^{\alpha}_{\beta}\ w^{\beta}\ \partial_{\alpha}$$

as in (9.18). It follows that $B^{\alpha}_{\beta} = b^{\alpha}_{\beta}$. By (9.7) we know that

$$B = i\ t\ 1_m + a$$

where $a \in H(m, \mathbb{R})$ and t is a real number. Therefore

$$(9.21) \qquad \begin{aligned} & b^{\alpha}_{\mu} \in \mathbb{R}\ (\alpha \ne \mu) \\ & \operatorname{Im} b^{\alpha}_{\alpha} = t \quad \text{(independent of } \alpha\text{)}. \end{aligned}$$

By the above formula for b^{α}_{γ}, the first relation means that if $\alpha \neq \gamma$

$$2\,\bar{c}^{\gamma} d^{\alpha}_{\gamma\gamma} + \sum_{\beta \neq \gamma} \bar{c}^{\beta}\, d^{\alpha}_{\beta\gamma} + c^{\alpha}\, \bar{d}^{\alpha}_{\alpha\gamma}$$

is purely imaginary. Put $c = (0,..,\overset{\delta}{1},..,0)$. Then (i) if $\delta = \gamma$, $d^{\alpha}_{\gamma\gamma}$ is purely imaginary; (ii) if $\delta = \alpha$, $d^{\alpha}_{\alpha\gamma} + \bar{d}^{\alpha}_{\alpha\gamma}$ is purely imaginary and, so $d^{\alpha}_{\alpha\gamma} = 0$; (iii) if $\delta \neq \alpha, \gamma$, then $d^{\alpha}_{\delta\gamma}$ is purely imaginary. These being true for any $X \in \mathfrak{g}_{\frac{1}{2}}$ by the principle explained before, $d^{\alpha}_{\delta\gamma} = 0$ in these cases and we have proved

(9.22) $$d^{\alpha}_{\gamma\gamma} = d^{\alpha}_{\alpha\gamma} = d^{\alpha}_{\delta\gamma} = 0$$

for any α, γ, δ which are different each other. Therefore, the expression of b^{α}_{α} obtained before reduces to

$$\begin{aligned} b^{\alpha}_{\alpha} &= 2i\,\bar{c}^{\alpha}\, d^{\alpha}_{\alpha\alpha} + 4i\, c^{\alpha}\,\bar{d}^{\alpha}_{\alpha\alpha} \\ &= 2i\, c^{\alpha}\, \bar{d}^{\alpha}_{\alpha\alpha} + 4i\, \mathrm{Re}(\bar{c}^{\alpha}\, d^{\alpha}_{\alpha\alpha}). \end{aligned}$$

Take the case where $c = (0,.,\overset{\alpha}{1},..,0)$. Then we see $b^{\beta}_{\beta} = 0$ for $\beta \neq \alpha$. Since $m > 1$, it follows from (9.21) that $\mathrm{Im}\, b^{\alpha}_{\alpha} = 0$. On the other hand

$$\begin{aligned} \mathrm{Im}\, b^{\alpha}_{\alpha} &= \mathrm{Im}\,(2i\,\bar{d}^{\alpha}_{\alpha\alpha} + 4i\, d^{\alpha}_{\alpha\alpha}) \\ &= 6\,\mathrm{Re}\, d^{\alpha}_{\alpha\alpha}. \end{aligned}$$

Therefore $d^{\alpha}_{\alpha\alpha}$ is purely imaginary. Again by the principle given before, it follows

(9.23) $$d^{\alpha}_{\alpha\alpha} = 0.$$

The formulae (9.22) and (9.23) show that $p^{\kappa}_{1,0} = 0$. Then, as remarked in the beginning of this proof, we have $X = 0$. Thus we have proved $\mathfrak{g}_{\frac{1}{2}} = (0)$ for the Case C II.

(Q.E.D.)

We shall now study the structure of the subspace $\mathfrak{g}_1$. According to Theorem 8.2 and since $\mathfrak{g}_{\frac{1}{2}} = (0)$, $\mathfrak{g}_1$ consists of vector fields

$$X = \sum_{k \leq \ell} p^{k\ell}_{2,0}\, \partial_{k\ell} + \sum_{\alpha} p^{\alpha}_{1,1}\, \partial_{\alpha} \tag{9.24}$$

such that

$$[Y, X] \in \mathfrak{g}_0 \qquad \text{for any} \quad Y \in \mathfrak{g}_{-1} \tag{9.25}$$

and

$$\operatorname{Im} \operatorname{Tr} [X, Y] = 0 \quad \text{for any} \quad Y \in \mathfrak{g}_{-1} \tag{9.26}$$

Put

$$p^{k\ell}_{2,0} = \sum_{\substack{k \leq \ell \\ h \leq j \\ r \leq s}} A^{k\ell}_{hj,\, rs}\, z^{hj} z^{rs} \qquad (A^{k\ell}_{hj,rs} = A^{k\ell}_{rs,hj}) \tag{9.27}$$

$$p^{\alpha}_{1,1} = \sum_{\substack{h \leq j \\ \alpha, \beta}} B^{\alpha}_{hj,\beta}\, z^{hj} w^{\beta} \tag{9.28}$$

Then

$$[\partial_{hj}, X] = 2 \sum_{\substack{r \leq s \\ k \leq \ell}} A^{k\ell}_{hj,rs}\, z^{rs} \partial_{k\ell} + \sum_{\alpha, \beta} B^{\alpha}_{hj,\beta} w^{\beta} \partial_{\alpha}$$

This belongs to $\mathfrak{g}_0$ and, if this element is represented by (A, B) under (9.7), then the matrix

$$A_{hj} = (2 A^{k\ell}_{hj,rs})$$

represents the linear transformation $\tilde{A} : z \longrightarrow Az + z\,{}^{t}A$ on $H(m + q, \mathbb{C})$ and the matrix

$$B_{hj} = (B^{\alpha}_{hj,\beta})$$

is equal to B. Since A is a real matrix of the form $\begin{pmatrix} a & b \\ o & c \end{pmatrix}$ and $B = a + i\,t\,1_m$ $(t \in \mathbb{R})$, the matrix A_{hj} is also real matrix and

$$\begin{aligned} \operatorname{Im} \operatorname{Tr} [\partial_{hj}, X] &= \operatorname{Im} (\operatorname{Tr} A_{hj} + \operatorname{Tr} B_{hj}) \\ &= m\,t\,. \end{aligned}$$

Thus the condition (9.26) implies $t = 0$ and $B_{hj} = a$. In particular, the second term of the expression (9.24) of X is determined by its first term.

Now, consider the associated tube domain in $\mathbb{C}^n$:

$$\begin{aligned} D' &= D \cap \left\{ H(m+q, \mathbb{C}) \times \{0\} \right\} \\ &= D\,(\,H^+\,(m+q, \mathbb{R}\,)). \end{aligned}$$

We remark that the domain $D = D(\Omega, F)$ is contained in $D' \times \mathbb{C}^n$ as follows from the definition of $D(\Omega, F)$. By Theorem 6.3, there exists a homomorphism $\mathfrak{J}$ of the Lie subalgebra $\mathfrak{g}_{-1} + \mathfrak{g}_0 + \mathfrak{g}_1$ into the Lie algebra $\mathfrak{g}(D')$ defined by restricting vector fields to D'. Let

$$\mathfrak{g}(D') = \mathfrak{g}'_{-1} + \mathfrak{g}'_0 + \mathfrak{g}'_1$$

be the decomposition of $\mathfrak{g}(D')$. Then $\mathfrak{J}$ maps $\mathfrak{g}_{-1}, \mathfrak{g}_0, \mathfrak{g}_1$ into $\mathfrak{g}'_{-1}, \mathfrak{g}'_0, \mathfrak{g}'_1$ respectively. The property derived from (9.26) shows that $\mathfrak{J}$ restricted to $\mathfrak{g}_1$ is an injection of $\mathfrak{g}_1$ into $\mathfrak{g}'_1$. On the other hand, since $\mathfrak{g}_{-1}$ has $\left\{ \partial_{k\ell} \mid k \leq \ell \right\}$ as a basis, we see that $\mathfrak{J}$ maps $\mathfrak{g}_{-1}$ onto $\mathfrak{g}'_{-1}$ bijectively. Therefore we have by (9.25)

(9.29) $\quad [\,Y', \mathfrak{J}(X)] \in \mathfrak{J}(\mathfrak{g}_0)$ for any $Y' \in \mathfrak{g}'_{-1}$ for a given $X \in \mathfrak{g}_1$.

Now, according to the Case A, we may put

$$(9.30) \qquad \mathfrak{g}(D') = \mathfrak{sp}(m+q, \mathbb{R})$$

$$= \left\{ \begin{pmatrix} A & B \\ C & D \end{pmatrix} \;\middle|\; \begin{matrix} B,\, C \quad H(m+q, \mathbb{R}) \\ D = -{}^{t}A \end{matrix} \right\}$$

and

$$\mathfrak{g}'_{-1} = \left\{ \begin{pmatrix} 0 & B \\ 0 & 0 \end{pmatrix} \right\}, \quad \mathfrak{g}'_{0} = \left\{ \begin{pmatrix} A & 0 \\ 0 & -{}^{t}A \end{pmatrix} \right\}$$

$$\mathfrak{g}'_{1} = \left\{ \begin{pmatrix} 0 & 0 \\ C & 0 \end{pmatrix} \right\}$$

It follows then from Proposition 9.1

$$(9.31) \qquad \zeta(\mathfrak{g}_0) = \left\{ \begin{pmatrix} A & 0 \\ 0 & -{}^{t}A \end{pmatrix} \;\middle|\; A = \begin{pmatrix} a & b \\ 0 & c \end{pmatrix} \right\}$$

We shall now examine $\zeta(\mathfrak{g}_0)$ in Cases C I and C II separately.

<u>Case C I</u>: $m > 0, \quad q > 0$

Let $X' = \zeta(X)$ for $X \in \mathfrak{g}_1$ and Y' be any element of $\mathfrak{g}'_{-1}$.

We may put

$$X' = \begin{pmatrix} 0 & 0 \\ C & 0 \end{pmatrix}, \quad Y' = \begin{pmatrix} 0 & B \\ 0 & 0 \end{pmatrix}$$

where $B, C \in H(m+q, \mathbb{R})$. Then

$$[Y', X'] = \begin{pmatrix} BC & 0 \\ 0 & -CB \end{pmatrix}.$$

The condtion (9.29) means that this belongs to $\zeta(\mathfrak{g}_0)$. By (9.31) it follows that $B\,C$ is of the form A in (9.31). This being true for any $B \in H(m+q, \mathbb{R})$, one sees easily that C is of the form

$$(9.32) \qquad C = \begin{pmatrix} 0 & 0 \\ 0 & c \end{pmatrix}$$

where $c \in H(q, \mathbb{R})$. We shall show that when X varies over $\mathfrak{g}_1$, c varies all over $H(q, \mathbb{R})$. Then, this will prove that $\mathfrak{g}_1$ may be identified with $H(q, \mathbb{R})$ through the map ξ.

To prove this assertion, recall first that the group $Sp(m+q, \mathbb{R})$ acts on the Siegel's upper half space $D' = D(H^+(m+q, \mathbb{R}))$ as explained in Case A. In particular, an element of the form

$$(9.33) \qquad \begin{pmatrix} 1_{m+q} & 0 \\ C & 1_{m+q} \end{pmatrix}$$

with $C \in H(m+q, \mathbb{C})$ belongs to $Sp(m+q, \mathbb{R})$ and transforms a point $z \in D'$ to the point $z(Cz + 1_{m+q})^{-1}$. Moreover, we get

$$(9.34) \qquad \operatorname{Im}(z\,(Cz + 1_{m+q})^{-1}) = {}^t\overline{(Cz + 1_{m+q})}^{-1}\,(\operatorname{im} z)\,(Cz + 1_{m+q})^{-1}.$$

Assume now that C is of the form

$$\begin{pmatrix} 0 & 0 \\ 0 & tc \end{pmatrix}$$

with $c \in H(q, \mathbb{R})$ and $t \in \mathbb{R}$. Then, since ${}^t\overline{(Cz + 1_{m+q})}^{-1}\,F(w,w)\,(Cz + 1_{m+q})^{-1} = F(w,w)$, it follows from (9.34), $(z\,(Cz + 1_{m+q})^{-1}, w) \in D$ when $(z,w) \in D$. In other words, if X is the element of the Lie algebra $\mathfrak{sp}(m+q, \mathbb{R})$ given by

$$X = \begin{pmatrix} 0 & 0 \\ C & 0 \end{pmatrix}$$

where C is the element in (9.32), then the one-parameter group of transformations of $D' \times \mathbb{C}^m$ given by

$$(z,w) \longrightarrow (\exp tX\, z, w)$$

preserves D. As we see easily, this one-parameter group defines a vector field $X \in \mathfrak{g}_1$ on D such that $\zeta(X) = X'$ is expressed with C in (9.32). Thus, any $c \in H(q, \mathbb{R})$ appears in C for some $X \in \mathfrak{g}_1$, which was to be proved.

<u>Case C II</u>: $m > 1$, $q = 0$.

We show that ζ maps $\mathfrak{g}_1$ onto $\mathfrak{g}_1'$ in this case. Let C be any element of $H(m, \mathbb{R})$. We define its action on $D' \times \mathbb{C}^m$ by

$$C(z,w) = (z', w')$$

where $z' = z\,(Cz + 1_m)^{-1}$ and $w' = {}^t\overline{(Cz + 1_m)}^{-1} w$. Then, applying (3.24) we get the following relation.

$$\begin{aligned} &\operatorname{Im} z' - F(w',w') \\ &= \tfrac{1}{2} \left\{ {}^t\overline{Q}\, (\operatorname{Im} z - F(w,w))\, Q + {}^tQ\, (\operatorname{Im} z - F(w,w))\, \overline{Q} \right\} \\ &\quad + F({}^tQ\, w, {}^tQ\, w) \end{aligned}$$

where we put $Q = (Cz + 1_m)^{-1}$. Since the right-hand side belongs to Ω for $(z,w) \in D$, it follows that the action of C on $D' \times \mathbb{C}^m$ preserves D. As in Case C I, we see then that any $C \in H(m, \mathbb{R})$ regarded as element of $\mathfrak{g}_1'$ is the image of some $X \in \mathfrak{g}_1$ under the map ζ. Thus $\mathfrak{g}_1$ may be identified with $H(m, \mathbb{R})$.

We have so far proved the following proposition.

Proposition 9.3. In Cases C I and C II, $\mathfrak{g}_1$ may be canonically identified with $H(q, \mathbb{R})$ and $H(m, \mathbb{R})$ respectively.

From the results obtained in the course of proving Proposition 9.1, we can see moreover the followings. In Case C I, the group $Sp(q, \mathbb{R})$ acting on D' as a subgroup of $Sp(m + q, \mathbb{R})$ and trivially on $\mathbb{C}^m$, the group acts on $D' \times \mathbb{C}^m$ preserving the domain D. In Case C II, the action of the group $Sp(m, \mathbb{R})$ on D' can be extended to an action on $D' \times \mathbb{C}^m$ preserving the domain D. Besides, applying Theorem 6.2, we have the following decompositions and isomorphisms, where $\mathfrak{r}$ and $\mathfrak{s}$ denote the radical and a maximal semisimple subalgebra of $\mathfrak{g}(D)$ respectively.

Case C I : $m > 0$, $q > 0$

$$\mathfrak{g}_{-1} = \mathfrak{r}_{-1} + \mathfrak{s}_{-1} \quad \text{(direct)},$$

$$\mathfrak{g}_{-1} \cong H(m + q, \mathbb{R}),$$

$$\mathfrak{r}_{-1} = H(m, \mathbb{R}) + M(m, q, \mathbb{R})$$

$$\mathfrak{s}_{-1} = H(q, \mathbb{R})$$

$$\mathfrak{g}_{-\frac{1}{2}} = \mathfrak{r}_{-\frac{1}{2}} \cong \mathbb{C}^m$$

$$\mathfrak{g}_0 = \mathfrak{r}_0 + \mathfrak{s}_0 \,, \quad \text{(direct)}$$

$$\mathfrak{r}_0 = M(m, q, \mathbb{R}) + \mathbb{C}$$

$$\mathfrak{s}_0 = \mathfrak{sl}(m, \mathbb{R}) + \mathfrak{gl}(q, \mathbb{R})$$

$$\mathfrak{g}_{\frac{1}{2}} = (0)$$

$$\mathfrak{g}_1 = \mathfrak{s}_1 = H(q, \mathbb{R})$$

$$\mathfrak{w} = \mathfrak{w}_{-1} + \mathfrak{w}_{-\frac{1}{2}} + \mathfrak{w}_0$$

$$\mathfrak{r} = \mathfrak{r}_{-1} + \mathfrak{r}_0 + \mathfrak{r}_1$$

$$= \mathfrak{gl}(m, \mathbb{R}) + \mathfrak{gz}(q, \mathbb{R})$$

<u>Case C II</u> : $m > 1$, $q = 0$.

$$\mathfrak{g}_{-1} = \mathfrak{r}_{-1} = H(m, \mathbb{R})$$

$$\mathfrak{g}_{-\frac{1}{2}} = \mathfrak{w}_{-\frac{1}{2}} = \mathbb{C}^m$$

$$\mathfrak{g}_0 = \mathfrak{w}_0 + \mathfrak{r}_0$$

$$\mathfrak{w}_0 = \mathbb{R}, \quad \mathfrak{r}_0 = \mathfrak{gl}(m, \mathbb{R})$$

$$\mathfrak{g}_{\frac{1}{2}} = (0)$$

$$\mathfrak{g}_1 = \mathfrak{r}_1 = H(m, \mathbb{R})$$

$$\mathfrak{w} = \mathfrak{w}_{-\frac{1}{2}} + \mathfrak{w}_0$$

$$\mathfrak{r} = \mathfrak{r}_{-1} + \mathfrak{r}_0 + \mathfrak{r}_1 = \mathfrak{sp}(m, \mathbb{R})$$

<u>Remark</u>. These examples were discussed by Tanaka [2] where, however, Case C II was not distinguished from Case C I. Takeuchi pointed out this error calculating these examples a new depending on results of Pjateckii-Sapiro. The Case C I with $m = q = 1$ is the example introduced at the end of § 3 : Since $\mathfrak{g}(D)$ is not semi-simple, the domain D is not symmetric for the Cases C I and C II.

Lecture Notes in Mathematics

Comprehensive leaflet on request

Vol. 111: K. H. Mayer, Relationen zwischen charakteristischen Zahlen. III, 99 Seiten. 1969. DM 16,-

Vol. 112: Colloquium on Methods of Optimization. Edited by N. N. Moiseev. IV, 293 pages. 1970. DM 18,-

Vol. 113: R. Wille, Kongruenzklassengeometrien. III, 99 Seiten. 1970. DM 16,-

Vol. 114: H. Jacquet and R. P. Langlands, Automorphic Forms on GL (2). VII, 548 pages. 1970. DM 24,

Vol. 115: K. H. Roggenkamp and V. Huber-Dyson, Lattices over Orders I. XIX, 290 pages. 1970. DM 18,-

Vol. 116: Séminaire Pierre Lelong (Analyse) Année 1969. IV, 195 pages. 1970. DM 16,-

Vol. 117: Y. Meyer, Nombres de Pisot, Nombres de Salem et Analyse Harmonique. 63 pages. 1970. DM 16,-

Vol. 118: Proceedings of the 15th Scandinavian Congress, Oslo 1968. Edited by K. E. Aubert and W. Ljunggren. IV, 162 pages. 1970. DM 16,-

Vol. 119: M. Raynaud, Faisceaux amples sur les schémas en groupes et les espaces homogènes. III, 219 pages. 1970. DM 16,-

Vol. 120: D. Siefkes, Büchi's Monadic Second Order Successor Arithmetic. XII, 130 Seiten. 1970. DM 16,-

Vol. 121: H. S. Bear, Lectures on Gleason Parts. III, 47 pages. 1970. DM 16,-

Vol. 122: H. Zieschang, E. Vogt und H.-D. Coldewey, Flächen und ebene diskontinuierliche Gruppen. VIII, 203 Seiten. 1970. DM 16,-

Vol. 123: A. V. Jategaonkar, Left Principal Ideal Rings. VI, 145 pages. 1970. DM 16,-

Vol. 124: Séminare de Probabilités IV. Edited by P. A. Meyer. IV, 282 pages. 1970. DM 20,-

Vol. 125: Symposium on Automatic Demonstration. V, 310 pages. 1970. DM 20,-

Vol. 126: P. Schapira, Théorie des Hyperfonctions. XI, 157 pages. 1970. DM 16,-

Vol. 127: I. Stewart, Lie Algebras. IV, 97 pages. 1970. DM 16,-

Vol. 128: M. Takesaki, Tomita's Theory of Modular Hilbert Algebras and its Applications. II, 123 pages. 1970. DM 16,-

Vol. 129: K. H. Hofmann, The Duality of Compact Semigroups and C*-Bigebras. XII, 142 pages. 1970. DM 16,-

Vol. 130: F. Lorenz, Quadratische Formen über Körpern. II, 77 Seiten. 1970. DM 16,-

Vol. 131: A Borel et al., Seminar on Algebraic Groups and Related Finite Groups. VII, 321 pages. 1970. DM 22,-

Vol. 132: Symposium on Optimization. III, 348 pages. 1970. DM 22,-

Vol. 133: F. Topsøe, Topology and Measure. XIV, 79 pages. 1970. DM 16,-

Vol. 134: L. Smith, Lectures on the Eilenberg-Moore Spectral Sequence. VII, 142 pages. 1970. DM 16,-

Vol. 135: W. Stoll, Value Distribution of Holomorphic Maps into Compact Complex Manifolds. II, 267 pages. 1970. DM 18,-

Vol. 136: M. Karoubi et al., Séminaire Heidelberg-Saarbrücken-Strasbourg sur la K-Théorie. IV, 264 pages. 1970. DM 18,-

Vol. 137: Reports of the Midwest Category Seminar IV. Edited by S. MacLane. III, 139 pages. 1970. DM 16,-

Vol. 138: D. Foata et M. Schützenberger, Théorie Géométrique des Polynômes Eulériens. V, 94 pages. 1970. DM 16,-

Vol. 139: A. Badrikian, Séminaire sur les Fonctions Aléatoires Linéaires et les Mesures Cylindriques. VII, 221 pages. 1970. DM 18,-

Vol. 140: Lectures in Modern Analysis and Applications II. Edited by C. T. Taam. VI, 119 pages. 1970. DM 16,-

Vol. 141: G. Jameson, Ordered Linear Spaces. XV, 194 pages. 1970. DM 16,-

Vol. 142: K. W. Roggenkamp, Lattices over Orders II. V, 388 pages. 1970. DM 22,-

Vol. 143: K. W. Gruenberg, Cohomological Topics in Group Theory. XIV, 275 pages. 1970. DM 20,-

Vol. 144: Seminar on Differential Equations and Dynamical Systems, II. Edited by J. A. Yorke. VIII, 268 pages. 1970. DM 20,-

Vol. 145: E. J. Dubuc, Kan Extensions in Enriched Category Theory. XVI, 173 pages. 1970. DM 16,-

Vol. 146: A. B. Altman and S. Kleiman, Introduction to Grothendieck Duality Theory. II, 192 pages. 1970. DM 18,-

Vol. 147: D. E. Dobbs, Cech Cohomological Dimensions for Commutative Rings. VI, 176 pages. 1970. DM 16,-

Vol. 148: R. Azencott, Espaces de Poisson des Groupes Localement Compacts. IX, 141 pages. 1970. DM 16,-

Vol. 149: R. G. Swan and E. G. Evans, K-Theory of Finite Groups and Orders. IV, 237 pages. 1970. DM 20,-

Vol. 150: Heyer, Dualität lokalkompakter Gruppen. XIII, 372 Seiten. 1970. DM 20,-

Vol. 151: M. Demazure et A. Grothendieck, Schémas en Groupes I. (SGA 3). XV, 562 pages. 1970. DM 24,-

Vol. 152: M. Demazure et A. Grothendieck, Schémas en Groupes II. (SGA 3). IX, 654 pages. 1970. DM 24,-

Vol. 153: M. Demazure et A. Grothendieck, Schémas en Groupes III. (SGA 3). VIII, 529 pages. 1970. DM 24,

Vol. 154: A. Lascoux et M. Berger, Variétés Kähleriennes Compactes. VII, 83 pages. 1970. DM 16,-

Vol. 155: Several Complex Variables I, Maryland 1970. Edited by J. Horváth. IV, 214 pages. 1970. DM 18,-

Vol. 156: R. Hartshorne, Ample Subvarieties of Algebraic Varieties. XIV, 256 pages. 1970. DM 20,-

Vol. 157: T. tom Dieck, K. H. Kamps und D. Puppe, Homotopietheorie. VI, 265 Seiten. 1970. DM 20,-

Vol. 158: T. G. Ostrom, Finite Translation Planes. IV. 112 pages. 1970. DM 16,-

Vol. 159: R. Ansorge und R. Hass. Konvergenz von Differenzenverfahren für lineare und nichtlineare Anfangswertaufgaben. VIII, 145 Seiten. 1970. DM 16,-

Vol. 160: L. Sucheston, Constributions to Ergodic Theory and Probability. VII, 277 pages. 1970. DM 20,-

Vol. 161: J. Stasheff, H-Spaces from a Homotopy Point of View. VI, 95 pages. 1970. DM 16,-

Vol. 162: Harish-Chandra and van Dijk, Harmonic Analysis on Reductive p-adic Groups. IV, 125 pages. 1970. DM 16,-

Vol. 163: P. Deligne, Equations Différentielles à Points Singuliers Reguliers. III, 133 pages. 1970. DM 16,-

Vol. 164: J. P. Ferrier, Seminaire sur les Algebres Complétes. II, 69 pages. 1970. DM 16,-

Vol. 165: J. M. Cohen, Stable Homotopy. V, 194 pages. 1970. DM 16.

Vol. 166: A. J. Silberger, PGL_2 over the p-adics: its Representations, Spherical Functions, and Fourier Analysis. VII, 202 pages. 1970. DM 18,-

Vol. 167: Lavrentiev, Romanov and Vasiliev, Multidimensional Inverse Problems for Differential Equations. V, 59 pages. 1970. DM 16,-

Vol. 168: F. P. Peterson, The Steenrod Algebra and its Applications: A conference to Celebrate N. E. Steenrod's Sixtieth Birthday. VII, 317 pages. 1970. DM 22,-

Vol. 169: M. Raynaud, Anneaux Locaux Henséliens. V, 129 pages. 1970. DM 16,-

Vol. 170: Lectures in Modern Analysis and Applications III. Edited by C. T. Taam. VI, 213 pages. 1970. DM 18,-

Vol. 171: Set-Valued Mappings, Selections and Topological Properties of 2^X. Edited by W. M. Fleischman. X, 110 pages. 1970. DM 16,-

Vol. 172: Y.-T. Siu and G. Trautmann, Gap-Sheaves and Extension of Coherent Analytic Subsheaves. V, 172 pages. 1971. DM 16,-

Vol. 173: J. N. Mordeson and B. Vinograde, Structure of Arbitrary Purely Inseparable Extension Fields. IV, 138 pages. 1970. DM 16,-

Vol. 174: B. Iversen, Linear Determinants with Applications to the Picard Scheme of a Family of Algebraic Curves. VI, 69 pages. 1970. DM 16,-

Vol. 175: M. Brelot, On Topologies and Boundaries in Potential Theory. VI, 176 pages. 1971. DM 18,-

Vol. 176: H. Popp, Fundamentalgruppen algebraischer Mannigfaltigkeiten. IV, 154 Seiten. 1970. DM 16,

Vol. 177: J. Lambek, Torsion Theories, Additive Semantics and Rings of Quotients. VI, 94 pages. 1971. DM 16,-

Please turn over

Vol. 178: Th. Bröcker und T. tom Dieck, Kobordismentheorie. XVI, 191 Seiten. 1970. DM 18,–

Vol. 179: Seminaire Bourbaki – vol. 1968/69. Exposés 347-363. IV, 295 pages. 1971. DM 22,–

Vol. 180: Séminaire Bourbaki – vol. 1969/70. Exposés 364-381. IV, 310 pages. 1971. DM 22,–

Vol. 181: F. DeMeyer and E. Ingraham, Separable Algebras over Commutative Rings. V, 157 pages. 1971. DM 16.

Vol. 182: L. D. Baumert. Cyclic Difference Sets. VI, 166 pages. 1971. DM 16,–

Vol. 183: Analytic Theory of Differential Equations. Edited by P. F. Hsieh and A. W. J. Stoddart. VI, 225 pages. 1971. DM 20,–

Vol. 184: Symposium on Several Complex Variables, Park City, Utah, 1970. Edited by R. M. Brooks. V, 234 pages. 1971. DM 20,–

Vol. 185: Several Complex Variables II, Maryland 1970. Edited by J. Horváth. III, 287 pages. 1971. DM 24,

Vol. 186: Recent Trends in Graph Theory. Edited by M. Capobianco/ J. B. Frechen/M. Krolik. VI, 219 pages. 1971. DM 18,–

Vol. 187: H. S. Shapiro, Topics in Approximation Theory. VIII, 275 pages. 1971. DM 22,–

Vol. 188: Symposium on Semantics of Algorithmic Languages. Edited by E. Engeler. VI, 372 pages. 1971. DM 26,–

Vol. 189: A. Weil, Dirichlet Series and Automorphic Forms. V, 164 pages. 1971. DM 16,–

Vol. 190: Martingales. A Report on a Meeting at Oberwolfach, May 17-23, 1970. Edited by H. Dinges. V, 75 pages. 1971. DM 16,–

Vol. 191: Séminaire de Probabilités V. Edited by P. A. Meyer. IV, 372 pages. 1971. DM 26,–

Vol. 192: Proceedings of Liverpool Singularities – Symposium I. Edited by C. T. C. Wall. V, 319 pages. 1971. DM 24,–

Vol. 193: Symposium on the Theory of Numerical Analysis. Edited by J. Ll. Morris. VI, 152 pages. 1971. DM 16,–

Vol. 194: M. Berger, P. Gauduchon et E. Mazet. Le Spectre d'une Variété Riemannienne. VII, 251 pages. 1971. DM 22,–

Vol. 195: Reports of the Midwest Category Seminar V. Edited by J.W. Gray and S. Mac Lane. III, 255 pages. 1971. DM 22,–

Vol. 196: H-spaces – Neuchâtel (Suisse)- Août 1970. Edited by F. Sigrist, V, 156 pages. 1971. DM 16,–

Vol. 197: Manifolds – Amsterdam 1970. Edited by N. H. Kuiper. V, 231 pages. 1971. DM 20,–

Vol. 198: M. Hervé, Analytic and Plurisubharmonic Functions in Finite and Infinite Dimensional Spaces. VI, 90 pages. 1971. DM 16,–

Vol. 199: Ch. J. Mozzochi, On the Pointwise Convergence of Fourier Series. VII, 87 pages. 1971. DM 16,–

Vol. 200: U. Neri, Singular Integrals. VII, 272 pages. 1971. DM 22,–

Vol. 201: J. H. van Lint, Coding Theory. VII, 136 pages. 1971. DM 16,–

Vol. 202: J. Benedetto, Harmonic Analysis on Totally Disconnected Sets. VIII, 261 pages. 1971. DM 22,–

Vol. 203: D. Knutson, Algebraic Spaces. VI, 261 pages. 1971. DM 22,–

Vol. 204: A. Zygmund, Intégrales Singulières. IV, 53 pages. 1971. DM 16,–

Vol. 205: Séminaire Pierre Lelong (Analyse) Année 1970. VI, 243 pages. 1971. DM 20,–

Vol. 206: Symposium on Differential Equations and Dynamical Systems. Edited by D. Chillingworth. XI, 173 pages. 1971. DM 16,–

Vol. 207: L. Bernstein, The Jacobi-Perron Algorithm – Its Theory and Application. IV, 161 pages. 1971. DM 16,–

Vol. 208: A. Grothendieck and J. P. Murre, The Tame Fundamental Group of a Formal Neighbourhood of a Divisor with Normal Crossings on a Scheme. VIII, 133 pages. 1971. DM 16,–

Vol. 209: Proceedings of Liverpool Singularities – Symposium II. Edited by C. T. C. Wall. V, 280 pages. 1971. DM 22,–

Vol. 210: M. Eichler, Projective Varieties and Modular Forms. III, 118 pages. 1971. DM 16,–

Vol. 211: Théorie des Matroïdes. Edité par C. P. Bruter. III, 108 pages. 1971. DM 16,–

Vol. 212: B. Scarpellini, Proof Theory and Intuitionistic Systems. VII, 291 pages. 1971. DM 24,–

Vol. 213: H. Hogbe-Nlend, Théorie des Bornologies et Applications. V, 168 pages. 1971. DM 18,–

Vol. 214: M. Smorodinsky, Ergodic Theory, Entropy. V, 64 pages. 1971. DM 16,–

Vol. 215: P. Antonelli, D. Burghelea and P. J. Kahn, The Concordance-Homotopy Groups of Geometric Automorphism Groups. X, 140 pages. 1971. DM 16,–

Vol. 216: H. Maaß, Siegel's Modular Forms and Dirichlet Series. VII, 328 pages. 1971. DM 20,–

Vol. 217: T. J. Jech, Lectures in Set Theory with Particular Emphasis on the Method of Forcing. V, 137 pages. 1971. DM 16,–

Vol. 218: C. P. Schnorr, Zufälligkeit und Wahrscheinlichkeit. IV, 212 Seiten 1971. DM 20,–

Vol. 219: N. L. Alling and N. Greenleaf, Foundations of the Theory of Klein Surfaces. IX, 117 pages. 1971. DM 16,–

Vol. 220: W. A. Coppel, Disconjugacy. V, 148 pages. 1971. DM 16,–

Vol. 221: P. Gabriel und F. Ulmer, Lokal präsentierbare Kategorien. V, 200 Seiten. 1971. DM 18,–

Vol. 222: C. Meghea, Compactification des Espaces Harmoniques. III, 108 pages. 1971. DM 16,–

Vol. 223: U. Felgner, Models of ZF-Set Theory. VI, 173 pages. 1971. DM 16,–

Vol. 224: Revètements Etales et Groupe Fondamental. (SGA 1). Dirigé par A. Grothendieck XXII, 447 pages. 1971. DM 30,–

Vol. 225: Théorie des Intersections et Théorème de Riemann-Roch. (SGA 6). Dirigé par P. Berthelot, A. Grothendieck et L. Illusie. XII, 700 pages. 1971. DM 40,–

Vol. 226: Seminar on Potential Theory, II. Edited by H. Bauer. IV, 170 pages. 1971. DM 18,–

Vol. 227: H. L. Montgomery, Topics in Multiplicative Number Theory. IX, 178 pages. 1971. DM 18,–

Vol. 228: Conference on Applications of Numerical Analysis. Edited by J. Ll. Morris. X, 358 pages. 1971. DM 26,–

Vol. 229: J. Väisälä, Lectures on n-Dimensional Quasiconformal Mappings. XIV, 144 pages. 1971. DM 16,–

Vol. 230: L. Waelbroeck, Topological Vector Spaces and Algebras. VII, 158 pages. 1971. DM 16,–

Vol. 231: H. Reiter, L^1-Algebras and Segal Algebras. XI, 113 pages. 1971. DM 16,–

Vol. 232: T. H. Ganelius, Tauberian Remainder Theorems. VI, 75 pages. 1971. DM 16,–

Vol. 233: C. P. Tsokos and W. J. Padgett. Random Integral Equations with Applications to Stochastic Systems. VII, 174 pages. 1971. DM 18,–

Vol. 234: A. Andreotti and W. Stoll. Analytic and Algebraic Dependence of Meromorphic Functions. III, 390 pages. 1971. DM 26,–

Vol. 235: Global Differentiable Dynamics. Edited by O. Hájek, A. J. Lohwater, and R. McCann. X, 140 pages. 1971. DM 16,–

Vol. 236: M. Barr, P. A. Grillet, and D. H. van Osdol. Exact Categories and Categories of Sheaves. VII, 239 pages. 1971. DM 20,–

Vol. 237: B. Stenström. Rings and Modules of Quotients. VII, 136 pages. 1971. DM 16,–

Vol. 238: Der kanonische Modul eines Cohen-Macaulay-Rings. Herausgegeben von Jürgen Herzog und Ernst Kunz. VI, 103 Seiten. 1971. DM 16,–

Vol. 239: L. Illusie, Complexe Cotangent et Déformations I. XV, 355 pages. 1971. DM 26,–

Vol. 240: A. Kerber, Representations of Permutation Groups I. VII, 192 pages. 1971. DM 18,–

Vol. 241: S. Kaneyuki, Homogeneous Bounded Domains and Siegel Domains. V, 89 pages. 1971. DM 16,–

Vol. 242: R. R. Coifman et G. Weiss, Analyse Harmonique Non-Commutative sur Certains Espaces. V, 160 pages. 1971. DM 16,–

Vol. 243: Japan-United States Seminar on Ordinary Differential and Functional Equations. Edited by M. Urabe. VIII, 332 pages. 1971. DM 26,–

Vol. 244: Séminaire Bourbaki – vol. 1970/71. Exposés 382–399. IV, 356 pages. 1971. DM 26,–

Vol. 245: D. E. Cohen, Groups of Cohomological Dimension One. V, 99 pages. 1972. DM 16,–